Springer Series in Materials Science

Volume 283

The Springer Series in Materials Science covers the complete spectrum of materials research and technology, including fundamental principles, physical properties, materials theory and design. Recognizing the increasing importance of materials science in future device technologies, the book titles in this series reflect the state-of-the-art in understanding and controlling the structure and properties of all important classes of materials.

More information about this series at http://www.springer.com/series/856

Hamideh Khanbareh · Vitaly Yu. Topolov ·
Christopher R. Bowen

Piezo-Particulate Composites

Manufacturing, Properties, Applications

With 60 Figures

Springer

Hamideh Khanbareh
Department of Mechanical Engineering
University of Bath
Bath, UK

Vitaly Yu. Topolov
Department of Physics
Southern Federal University
Rostov-on-Don, Russia

Christopher R. Bowen
Department of Mechanical Engineering
University of Bath
Bath, UK

ISSN 0933-033X ISSN 2196-2812 (electronic)
Springer Series in Materials Science
ISBN 978-3-030-19203-7 ISBN 978-3-030-19204-4 (eBook)
https://doi.org/10.1007/978-3-030-19204-4

This Springer imprint is published by the registered company Springer Nature Switzerland AG
The registered company address is: Gewerbestrasse 11, 6330 Cham, Switzerland

To our friends, pupils and colleagues in the vast piezoelectric field

Preface

Broadly defined, composites are materials with two or more distinct components that combine to yield characteristics superior to those of the individual constituents.

The term 'composite' more specifically refers to a structural material (such as plastic) within which a fibrous material (such as silicon carbide) is embedded.

Encyclopædia Britannica

In the last decade smart materials and multifunctional systems [1] with a wide range of performance-tailored functionality are the focus of attention of specialists in the field of materials science, physics of functional materials, mechanics of heterogeneous media, and branches of fundamental sciences. Researchers are seeking to design highly effective materials that are able to sense, diagnose and respond to external stimuli or fields without any human intervention. Adaptive structures that allow a readjustment of functionality, shape and properties on demand, as well as self-sustaining systems for energy-harvesting applications are also of interest.

Multifunctional materials designed to improve system performance are often manufactured in the form of composites that consist of several distinct components, and each component of the composite performs an important function or functions [2, 3]. Composites can be designed to meet specific requirements through tailored properties, and these properties are effective (average) due to the heterogeneous character of the materials. The properties in a composite system can be integrated and combined at several dimensional scales, with increasing inter-connectivity between the individual components as the dimensional scale decreases.

A process of material selection is necessary to satisfy the requirements of achieving more than one function, for multifunctional materials and tailored structures. An important group of materials studied for ease of integration with others to create novel multifunctional materials are stimulus-responsive (smart) materials that react to a specific stimulus by a change in their physical properties [4]. Among these types of materials, poled ferroelectric ceramics (piezoceramics) have gained significant attention for a wide variety of applications in piezoelectric sensing, actuation, and energy harvesting; the areas of these applications has continued to broaden in recent decades. However, many ferroelectric ceramics have deficiencies such as low toughness, composition fluctuations and lack of formability. Particulate ferroelectric ceramics have been combined with inert or piezoelectric polymers [5] to form flexible functional composites with predictable effective properties. In such composites their functionalities are controlled by the intrinsic properties of the constituent ceramic and polymer components, their morphology, connectivity patterns, and technological factors. New developments in polymer synthesis have led to a variety of advanced flexible composites with functionalities in multiple domains of piezo- and pyroelectric in a wide temperature range. Moreover, the micro/macro-architecture design and optimisation of specific parameters by using a range of polymer processing techniques has resulted in the enhanced performance of the composites suitable for piezoelectric sensor and actuator applications.

In the present monograph we consider examples of modern piezo-particluate composites based on ferroelectric ceramics and discuss a wide cycle of problems on manufacturing, experimental studies and prediction of the effective physical properties. We pay specific attention to the piezoelectric performance of the ceramic/polymer composites that are to be taken into account for a variety of piezotechnical applications. We analyse numerous examples of the 'microgeometry – properties' relations in the piezo-particluate composites. Undoubtedly, this analysis widens the traditional materials-science concepts on the *composition—structure—properties* relations in the fundamental triangle [3, 6] and may be a stimulus to form novel highly effective piezo-active composite with predictable parameters.

The novelty of the present monograph consists in the systematisation of many authors' results on the manufacture and performance of modern piezo-particulate composites as well in the description of links between this performance and applications of these materials.

The present monograph has been written on the basis of the authors' research results obtained at the Delft University of Technology (The Netherlands), Southern Federal University (Russia) and University of Bath (UK). The academic style of the description of the research results and discussions about them in the context of manufacturing, properties and applications indicate that this monograph would be useful to engineers, postgraduate students, researchers, and lecturers which work in the field of piezoelectric and smart materials – from manufacturing to their effective properties. This monograph will be also useful to specialists working in the field of specific applications of novel active dielectrics and smart materials. We expect that some sections of the monograph can be a considered as a basis for a university course devoted to modern piezoelectric materials and their piezotechnical applications.

Introducing this new monograph, we would like to mention the following proverbs.

As you sow, you shall mow.
Experience is the mother of knowledge.
A tree is known by its fruit.
Many hands make light work.
Alone we can do so little; together we can do so much.

It is believed that the aforementioned proverbs can remind a reader some stages concerned with the manufacture of piezo-active composites, with the experimental study and modelling of their properties.

Finishing this Preface, we hope that the 21st century termed *The Century of New Materials and Technologies* will promote creation of novel highly effective ferroelectric, piezoelectric and related materials for various applications.

Bath, UK — Dr. Hamideh Khanbareh
Rostov-on-Don, Russia — Prof. Dr. Vitaly Yu. Topolov
Bath, UK — Prof. Dr. Christopher R. Bowen

References

1. S.A. Wilson, R.P.J. Jourdain, Q. Zhang, R.A. Dorey, C.R. Bowen, M. Willander, Q. Ul Wahab, M. Willander, S.M. Al-hilli, O. Nur, E. Quandt, C. Johansson, E. Pagounis, M. Kohl, J. Matovic, B. Samel, W. van der Wijngaart, E.W.H. Jager, D. Carlsson, Z. Djinovic, M. Wegener, C. Moldovan, R. Iosub, E. Abad, M. Wendlandt, C. Rusu, K. Persson, New materials for micro-scale sensors and actuators. An engineering review. Mater. Sci. Eng. R **56**, 1–129 (2017)
2. E.K. Akdogan, M. Allahverdi, A. Safari, Piezoelectric composites for sensor and actuator applications. IEEE Trans. Ultrason. Ferroelectr. Freq. Control **52**, 746–775 (2005)
3. V.Yu. Topolov, C.R. Bowen, *Electromechanical Properties in Composites Based on Ferroelectrics* (Springer, London, 2009)
4. R.E. Newnham, Molecular mechanisms in smart materials. Mater. Res. Soc. Bull. **22**(5), 20–34 (1997)
5. C.P. Bowen, R.E. Newnham, C.A. Randall, Dielectric properties of dielectrophoretically assembled particulate-polymer composites. J. Mater. Res. **13**, 205–210 (1998)
6. R.E. Newnham, *Properties of Materials. Anisotropy, Symmetry, Structure* (Oxford University Press, New York, 2005)

Acknowledgements

The authors are grateful to Prof. Dr. A. V. Turik, Prof. Dr. A. E. Panich, Prof. Dr. I. A. Parinov, Prof. Dr. A. A. Panich, Prof. Dr. A. A. Nesterov, Dr. V. K. Dolya, Dr. S. E. Filippov, and Dr. P. A. Borzov (Southern Federal University, Russia), Prof. Dr. J. T. Taylor and Dr. J. I. Roscow (University of Bath, UK), Prof. Dr. P. Bisegna (University of Rome Tor Vergata, Italy), Prof. Dr. O. Kraft and Prof. Dr. M. Kamlah (Karlsruhe Institute of Technology, Germany), Prof. Dr. M. Lethiecq and Prof. Dr. F. Levassort (University of Tours, France), Prof. Dr. S.-H. Chang (National Kaohsiung Marine University, Taiwan, Republic of China), Prof. Dr. W. Cao (The Pennsylvania State University, USA), Prof. Dr. A. Safari and Dr. E. K. Akdogan (Rutgers—The State University of New Jersey, USA), Prof. Dr. A. S. Sidorkin (Voronezh State University, Russia), and Prof. Dr. L. N. Korotkov and Prof. Dr. S. A. Gridnev (Voronezh State Technical University, Russia) for their interest in the research problems and for their research results that have been taken into account at writing the present monograph. The authors emphasise the vast geographic area wherein the piezo-active and related composites are studied, developed and manufactured. The author Prof. Dr. C. R. Bowen would also like to thank his past supervisors including Prof. Dr. B. Derby (University of Manchester, UK), Prof. Dr. N. Claussen (Technical University of Hamburg-Harburg, Germany) and Prof. Dr. R. Stevens (University of Bath, UK, who passed away in 2015 and is greatly thanked for support and missed). The author Dr. H. Khanbareh would like to thank her past supervisors including Prof. Dr. S. van der Zwaag and Prof. Dr. P. Groen (Delft University of Technology, the Netherlands).

The authors express their thanks to Prof. Dr. C. Ascheron and Mrs. E. Sauer (Springer-Verlag, Heidelberg, Germany) for their effective and timely co-operation in the field of editing and producing this monograph. The authors sincerely thank Mrs. C. A. Rothwell (University of Bath, UK) for the technical help. Copyright permissions obtained from Springer (www.springer.com), Elsevier (www.elsevier.com), IOP Publishing (www.iop.org), AIP (www.aip.org), Taylor & Francis (www.informaworld.com), Sage Publishing (sagepub.co.uk), Royal Society of Chemistry (www.rsc.org), and Prof. Dr. P. Bisegna (University of Rome Tor Vergata, Italy) are acknowledged with due attention and gratitude.

Financial support that promoted the fruitful research collaboration and writing this book is acknowledged with many thanks. Hereupon gratefully and proudly the authors mention the timely and effective support from the EPSRC (UK), National Physical Laboratory, QinetiQ (UK), Great Western Research (GWR, UK), University of Bath (UK), and Southern Federal University (Russia). The author Prof. Dr. C. R. Bowen gratefully acknowledges support of ERC Advanced Fellowship (Project "NEMESIS", Grant Agreement No. 320963).

Research has been carried out at the financial support from the Ministry of Education and Science of the Russian Federation within the framework of the complex project "Working out and creation of a high-technological production of a mobile hydroacoustic complex to highlight a situation in various areas of the World ocean on the basis of modern piezoelectric means of the new generation" (contract No. 03.G25.31.0276, May 29th, 2017) by using the equipment of the Centre of Collective Use "High Technologies" at the Southern Federal University. Hereby the author Prof. Dr. V. Yu. Topolov acknowledges support with thanks.

Bath, UK — Dr. Hamideh Khanbareh
Rostov-on-Don, Russia — Prof. Dr. Vitaly Yu. Topolov
Bath, UK — Prof. Dr. Christopher R. Bowen

Contents

About the Authors

Hamideh Khanbareh was born on August 9th, 1986 in Tehran, Iran. She received her B.Sc. degree, in 2008, in Materials Science and Engineering from University of Tehran. In 2012 she received her Cumlaude M.Sc. degree in Aerospace Engineering from Delft University of Technology, Netherlands. Her final project concerned fractal analysis of microstructures of ultra-high strength Aluminium alloys for aerospace applications. She then pursued her Ph.D. in Novel Aerospace Materials group, at Delft University of Technology working on functionally graded ferroelectric ceramic-polymer composites. During her Ph.D. she also worked as a visiting scientist at the Molecular Electronics Research Group at Max Planck Institute for Polymers (MPIP), in Mainz, Germany. In June 2016 she obtained her Ph.D. degree from Delft University of Technology and subsequently was appointed as a Prize Fellow at the Materials and Structures Research Centre, within the Department of Mechanical Engineering at University of Bath, UK, in September 2016. Dr. Khanbareh's main research interests are in materials design, modelling, fabrication and application of piezo- and pyroelectrics in sensing and energy harvesting. Heterogeneous polymer-ceramic composites, offering a wide range of compositional and microstructural design flexibility, are the target of her current research. Dr. H. Khanbareh has a strong record of publications, authoring over 20 peer-reviewed journal papers and 10 conference papers. She has been a member of IOM3 Smart Materials & Systems Committee (SMASC), Institute of Electrical and Electronics Engineers (IEEE) Ferroelectrics as well as Royal Society of Chemistry and UK Society of Biomaterials.

Vitaly Yu. Topolov was born in Rostov-on-Don, Russia (former USSR) on November 8th, 1961. He earned the qualification "Physicist. Educator" (honours degree) in 1984 along with the degrees "Candidate of Sciences (Physics and Mathematics)" and "Doctor of Sciences (Physics and Mathematics)" in 1987 and 2000, respectively, all from the Rostov State University, Russia. From 1987 to 1991, he worked as a research scientist of the Institute of Physics at the Rostov State University. From 1991 to 2000, he worked as a Senior Lecturer (1991–1992) and an Associate Professor (1992–2000) of the Department of Physics at the Rostov

State University. Since 2000, he has been a Professor of the same Department. Since December 2006, after reforming the Rostov State University, he is a Professor of the Department of Physics at the Southern Federal University (Rostov-on-Don, Russia).

Dr. V. Yu. Topolov was also a visiting scientist at the Moscow State University, Russia (former USSR, 1989), University of Saarland, Germany (1994–1995), Aachen University of Technology—RWTH Aachen, Germany (1998), Karlsruhe Research Center, Germany (2002 and 2003–2004), University of Bath, UK (2006, 2007, 2012, 2013, 2014, 2015, 2016, 2017, and 2018), and University of Rome Tor Vergata, Italy (2008). His research interests include heterogeneous ferroelectrics, smart materials, domain and heterophase structures, as well as electromechanical effects in ferroelectrics and related materials. He earned the special award from the International Science Foundation (1993) and the Soros Associate Professor title and awards from the International Soros Science-Educational Program and the Open Society Institute (1997, 1998, 2000, and 2001). He presented the best poster at the International Symposium on Ferroelectric Domains (China, 2000), the best oral report at the International Conference on Relaxation Phenomena in Solids (Russia, 2010), the best training aids in the Department of Physics, Southern Federal University (Russia, 2006 and 2011), and the best research works in the Department of High Technologies, Southern Federal University (Russia, 2008 and 2011). He is an author of six monographs published at *Springer* (London, UK, 2009; Berlin, Heidelberg, Germany, 2012; Berlin, Heidelberg, Germany, 2014; Cham, Switzerland, 2016 and 2018), two edited monographs published at *Springer* (Cham, Heidelberg, New York, Switzerland, Germany, USA, 2014 and 2016), one edited monograph published at *Nova Science Publishers* (New York, USA, 2016), five chapters in monographs published at *Nova Science Publishers* (New York, USA, 2010–2013), and over 450 scientific papers, reviews, conference proceedings, and abstracts. Dr. V. Yu. Topolov has been included in the list of Active Russian Scientists.

Christopher R. Bowen was born on January 18th, 1968 and grew up in Beddau, South Wales (UK). He earned a B.Sc. (First Class) in Materials Science at the School of Materials, University of Bath, UK in 1990 and worked on his D.Phil. thesis in ceramic manufacture under the supervision of Prof. Brian Derby in the Department of Materials, University of Oxford, UK in 1990–1993 (Ph.D. awarded in 1994). In 1993–1994, he worked as a Researcher in the Advanced Ceramics Group at the Technical University of Hamburg-Harburg (TUHH), Germany under the supervision of Prof. Nils Claussen. From 1994 to 1996, he was research fellow at the School of Materials, University of Leeds, UK working with Prof. Ron Stevens. From 1996 to 1998, he was a Senior Scientist at the Defence Evaluation and Research Agency (DERA), Functional Materials Group, UK working on ferroelectric ceramics and composites. He joined the University of Bath, UK in August 1998, and now he is a Professor at the same University.

The research interests of Dr. C. R. Bowen are concerned with modern functional materials, including ferroelectric ceramics and piezo-active composites for modern sensors and actuators, as well as with manufacturing and characterisation of these materials. Continuing interest includes the use of piezoelectric materials combined with structural composites, such as bistable laminates, for energy harvesting and shape changing applications. Additional interest is concerned with pyroelectric materials, their performance and use in energy-harvesting applications. Dr. C. R. Bowen earned the Thornton and Hazelwood prizes for academic work (1986–1990), Institute of Materials National Lecture Competition award (1993), SET award (2002), and John Willis award for excellence in research and teaching (2003). Dr. C. R. Bowen has published four monographs at *Springer* (London, UK, 2009; Berlin, Heidelberg, Germany, 2014; Cham, Switzerland, 2016 and 2018) and approximately 300 scientific papers, conference proceedings, and abstracts. He has been awarded an ERC Advanced Fellowship in Novel Energy Materials, Engineering Science and Integrated Systems (NEMESIS, Grant Agreement No. 320963).

Chapter 1
Piezo-Active Composites: Classification and Effective Physical Properties

Abstract Piezo-active composites are regarded as an important group of piezoelectric materials that belong to heterogeneous dielectrics whose physical properties and related parameters depend on many factors. Among the variety of criteria for the classification of piezo-active composites, their microgeometry and connectivity play an important role. Knowledge of the connectivity pattern enables us to carry out a prediction of the effective physical properties and related parameters in terms of micromechanical methods. Piezo-active composites play a key role among modern functional materials due to the considerable electromechanical coupling, piezoelectric activity, sensitivity and anisotropy, and figures of merit. This is achieved in the presence of highly effective ferroelectric components such as poled ferroelectric ceramics and domain-engineered relaxor-ferroelectric single crystals. The properties of the composites are regarded as effective properties in accordance with features of the microstructure, domain structure, arrangement of components and connectivity. The remarkable piezoelectric properties and related parameters of the composites based on ferroelectrics stimulate the creation of novel highly effective materials and are to be taken into account for potential transducer, hydroacoustic, energy-harvesting and other applications.

1.1 Piezo-Active Composites as Modern Active Dielectrics

The majority of modern dielectrics are heterogeneous materials, and among them ceramics, polycrystalline films and composites are characterised by various properties that are important for applications [1]. Composites are heterogeneous systems that consist of two or more components, and these components differ in chemical composition and properties and are separated by interfaces [2, 3]. Each composite material is characterised by the two following features [3, 4]:

(i) typical sizes of the separate structural elements, that form the component, are small in comparison with the whole composite sample, and
(ii) typical sizes of the structural elements are larger than the size of their individual atoms or molecules, and in this case each component is regarded as a continuous medium.

H. Khanbareh et al., *Piezo-Particulate Composites*, Springer Series in Materials Science 283, https://doi.org/10.1007/978-3-030-19204-4_1

Each component of the composite represents a set of the structural elements with equal physical constants. These elements are distributed either regularly (uniformly) or irregularly and can have a specific, random or variable shape. The composite can be described using a set of microgeometric characteristics such as connectivity, volume fractions of the components, and their spatial distribution [2–5].

The piezo-active composites can be regarded as an important group of piezoelectric materials that belong to heterogeneous dielectrics. This group of dielectrics is vast due to the large number of components that may be involved for the design of the piezo-active composite wherein at least one of the components is piezoelectric. Among the piezoelectric components, ferroelectric materials in the poled state play the leading role [1–3, 6–10]. Modern processing technologies enable the synthesis of a variety of ferroelectric single crystals, ferroelectric ceramics (FCs, polycrystalline materials), ferroelectric films, ferroelectric polymers and nano-sized ferroelectric materials. Ferroelectric properties are generally related to altering the spontaneous polarisation vector in an electric field (domain switching), and such a process is observed in acentric (non-centrosymmetric) crystal structures [11, 12]. Piezoelectric properties are concerned with an appearance of positive and negative charges on surfaces of a mechanically stressed sample [12, 13]. These properties are detected, for instance, in dielectric single crystals belonging to the non-centrosymmetric classes, and these classes are divided into two groups [14]. The first group contains ten symmetry classes (1, 2, 3, 4, 6, *m*, *mm*2, 3*m*, 4*mm*, and 6*mm*), and in these classes there is a singular polar axis that makes the single crystal polar. The second group contains 11 symmetry classes as follows: 222, $\bar{4}$, 422, $\bar{4}2m$, 32, $\bar{6}$, 622, $\bar{6}2m$, 23, $\bar{4}3m$, and 432. However, in the 432 class of the cubic system, the piezoelectric coefficients equal zero because of symmetry [14]. Thus, one can observe the piezoelectric effect in dielectric single crystals from the 20 symmetry classes. The piezoelectric effect can also be observed in acentric anisotropic dielectrics that are piezoelectric textures belonging to one of the following Curie groups [12–14]: ∞, ∞mm or $\infty/2$. Poled FCs exhibit piezoelectricity as piezoelectric textures with ∞mm symmetry, and the matrix of piezoelectric coefficients for this Curie group has the form of the matrix from 4*mm* to 6*mm* classes [12, 13].

In the last decades, perovskite-type compositions of FCs have been of interest due to their important physical properties, high electromechanical coupling, large piezoelectric anisotropy, high thermal stability, etc. [3, 11–13, 15]. FCs have been widespread piezoelectric components of various FC /polymer composites [2, 3, 6–9]. The piezoelectric properties of the perovskite-type and other FCs are highly dependent on manufacturing and poling methods, chemical composition, thermal treatment, mechanical stress history and microstructure [11, 15]. However, the FC materials are brittle and difficult to shape, and they possess a large acoustic impedance [16] that results in a more difficult acoustic match to soft media such as tissues and water.

In the perovskite-type FCs, the longitudinal piezoelectric coefficient (that characterises a charge at the sample surface per unit force applied along the normal vector to the sample) is $d_{33} \sim 10^2$ pC/N [11], and this value is two orders-of-magnitude higher than the piezoelectric coefficients d_{ij} of quartz ($d_{11} = -2.25$ pC/N and $d_{14} = 0.853$ pC/N [12]). As is well-known, quartz is a piezoelectric (non-ferroelectric, non-polar) single crystal on which the piezoelectric effect was discovered in 1880 [12–14]. In domain-engineered $(1-x)Pb(A_{1/3}Nb_{2/3})O_3 - xPbTiO_3$ single crystals with compositions chosen near the morphotropic phase boundary and A = Mg or Zn, typical experimental values are $d_{33} \sim 10^3$ pC/N [17, 18], i.e. about three orders-of-magnitude larger than the piezoelectric coefficients of quartz and one order-of-magnitude larger than the piezoelectric coefficients of conventional poled FCs.

Piezo-active composites based on ferroelectrics are of great due to possibilities to vary the microgeometry, effective physical properties and their anisotropy across a wide range. These variations appear in two-component composites, for instance, upon combining their components [2, 3, 7, 8] such as

(i) FC and polymer (that can be either piezoelectric or piezo-passive),
(ii) ferroelectric single crystal and FC, and
(iii) ferroelectric single crystal and polymer.

The presence of one or more piezoelectric components enables a set of composite parameters which are not particular to the separate components, and this synergy is important for many piezo-technical applications [1, 9, 16].

Piezoelectricity is an interaction between elastic variables (mechanical stress σ_{kl} and mechanical strain ξ_{qr}) and electric variables (electric field $\boldsymbol{E}$ and electric displacement $\boldsymbol{D}$) [12–14], and this interaction is of importance in homogeneous piezoelectric materials including composites. In the piezo-active composites, an electromechanical interaction between their components strongly depends on microgeometry and boundary conditions. As is known, the piezo-active composites based on ferroelectrics form the final link in hierarchy-of-properties chains of 'single-domain single crystal → polydomain single crystal → FC → composite' and 'single-domain single crystal → polydomain single crystal → heterophase single crystal → composite' [2]. The links of these chains can be studied by taking into account the properties of components and microgeometric features of the piezo-active composite.

1.2 Criteria of Classifications of Composites

Structure criteria play the important role in the classification of composites including piezo-active materials. In the last decades, the following criteria have been chosen to classify the composite materials [2, 3]:

(i) connectivity of the structural elements,
(ii) typical relative sizes and shape of structural elements and
(iii) features of the arrangement of structural elements.

The connectivity of the structural elements enables us to form a microgeometric image of the piezo-active composite to be studied. Based on features of the connectivity, we can divide the composites into three large groups [2, 3]. The first group is formed by so-called *matrix composites* wherein the structural elements of the first component represent a monolithic system (the matrix), and the structural elements of the remaining components are isolated and do not contact each other at small volume fractions. The second group comprises composites wherein each component is regarded as a monolithic and interconnected framework. The corresponding composite structures are *interpenetrated.* In the third group, the structural elements of each component of the composite are not interpenetrated or isolated.

Taking into consideration sizes and shapes of the structural elements, we define granular, fibrous and laminated composites [2, 3]. The granular composites consist of the structural elements having linear dimensions l_i of one order-of-magnitude in three co-ordinate directions ($i = 1, 2$ or 3). If $l_1 \approx l_2 \ll l_3$ in one of the components, then the composites are fibrous. In laminated composites, a relation $l_1 \approx l_2 \gg l_3$ holds for one or more components. If the latter relation is not true for all the components, such a composite can be regarded as a matrix composite reinforced by plate-like inclusions of one or more components.

Taking into account the arrangement of the structural elements, we analyse composites as materials with a *regular* (determined) or *irregular* (stochastic) structure [2, 3]. The structural elements in the composite with a regular structure are characterised by a relevant geometric shape (e.g. rectangular parallelepiped or ellipsoid) and a regular (equidistant and uniform) arrangement in the sample. At the irregular structure, the arrangement of the structural elements in the composite sample is random, and the shape of these elements can be either appointed or arbitrary. The random character of the arrangement can change and influence connectivity of the structural elements in the sample. An additional factor that characterises the arrangement of the structural elements is related to their spatial orientation.

1.3 Microgeometry and Connectivity

An important classification of two-component composites with planar interfaces was first put forward by Newnham et al. [6]. This classification takes into account microgeometric features and is based on the concept of the connectivity of each component in the composite. According to work [6], connectivity is expressed by the numbers of dimensions (or co-ordinate axes) in which each component is continuously distributed between limiting surfaces of the composite sample. The distribution of a self-connected state of a component can take place along zero, one, two or three co-ordinate axes, i.e. connectivity $\alpha = 0, 1, 2$ or 3 for the component 1 and connectivity $\beta = 0, 1, 2$ or 3 for the component 2. The connectivity of a two-component composite is written as α–β [1, 2, 5, 6] where the connectivity of the piezoelectric or most piezo-active component takes the first position (α). If $\alpha \leq \beta$, then the n-component composites are described by $(n + 3)!/(3!\, n!)$ connectivities [6], for $n = 2$ the number of connectivities is 10. The 10 connectivities at $\alpha \geq \beta$ are represented in Table 1.1. Table 1.1 contains the schematics of the two-component composites with systems of planar interfaces that separate the components.

Table 1.1 Examples of α–β connectivity patterns in composites with planar microgeometry

		First component			
Second compo-nent		$\alpha = 0$	$\alpha = 1$	$\alpha = 2$	$\alpha = 3$
	$\beta = 0$		Like a pattern at $\alpha = 0$ and $\beta = 1$	Like a pattern at $\alpha = 0$ and $\beta = 2$	Like a pattern at $\alpha = 0$ and $\beta = 3$
	$\beta = 1$	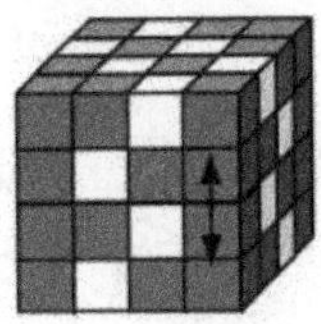	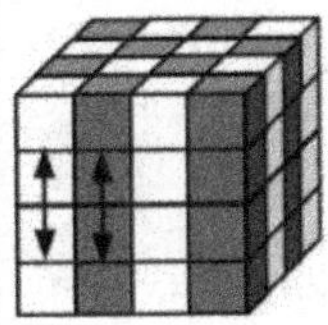		Like a pattern at $\alpha = 1$ and $\beta = 3$
	$\beta = 2$		Like a pattern at $\alpha = 2$ and $\beta = 1$		Like a pattern at $\alpha = 2$ and $\beta = 3$
	$\beta = 3$		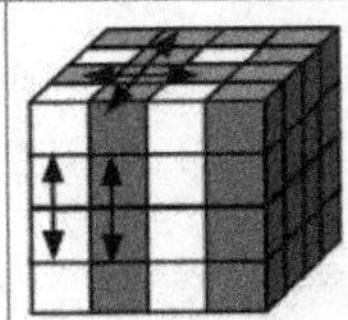		

Arrows show co-ordinate directions along which the first and second components are distributed continuously

In a general case, a two-component composite can be described in terms of $4^2 = 16$ connectivities: 0–0, 0–1, 0–2, 0–3, …, 3–2 and 3–3. The 6 connectivity patterns, which are not shown graphically in Table 1.1, can be obtained from the related patterns by a permutation of the components. For instance, the 3–1 connectivity pattern is similar to the 1–3 connectivity pattern, the 3–2 connectivity pattern is similar to the 2–3 connectivity pattern, etc.

If the composite contains three components, then its connectivity is denoted as α–β–γ, where α, β, γ = 0, 1, 2 or 3. As described earlier, α is related to the component with the highest piezoelectric activity. The second position is denoted by β and is related to the component that can be either piezo-active or piezo-passive and plays the important role in forming the effective properties of the composite as a whole.

The concept of connectivity [5, 6] is fundamental in developing an understanding of the electromechanical interaction between components within piezo-active composites, in the study of the distribution of internal electric and mechanical fields and in the interpretation of experimental or calculated data related to composites with specific connectivity [2, 7, 8]. The connectivity of piezo-active composites is crucial in influencing their piezoelectric properties and electromechanical coupling. Based on the concept of connectivity, piezo-active composites with the 0–3, 1–3, 2–2, 2–3, 3–1, 3–2 and 3–3 connectivity patterns have been experimentally studied (see, for instance, work [2, 9, 19, 20]).

An analogy between some composite structures shown in Table 1.1 is concerned with an evolution of the connectivity patterns [2]. For example, in a 0–3 FC/polymer composite, a system of isolated FC inclusions is surrounded by a large polymer matrix. A continuous distribution of the FC component along one of the co-ordinate axes (or a formation of FC chains, fibres or rods due to merging many inclusions) leads to 1–3 connectivity of the composite, and its effective properties will differ to the properties of the composite structure that has an initial 0–3 connectivity pattern.

As follows from work [2, 21], some connectivity patterns of two-component composite can be associated with small modifications of the initial connectivity pattern. Thus, there exists a group of so-called *junction* connectivity patterns whose combinations can form new connectivity patterns. The presence of a system of planar interfaces between the components in the α–β composite suggests that the 2–2 connectivity pattern (i.e. a system of series or parallel layers) can play an important role of one of the junction connectivity patterns. A simple multiplication of the 2–2 connectivity pattern results in the 1–1 connectivity pattern, and a simplification of the 1–1 connectivity pattern leads to the 1–3 or 3–1 connectivity pattern.

The main links between the connectivity patterns in two-component composites with planar interfaces are shown in Fig. 1.1. As follows from results [21], the α–β composite structures can be generated by systems of the layers that are either homogeneous (see C1 and C2 at the base of Fig. 1.1) or heterogeneous. The layers are formed by means of the junction connectivity patterns, sometimes by a combination of the junction connectivity pattern and one of the components. We add that knowledge of the evolution of the connectivity patterns enables an analysis of the effective electromechanical properties of the α–β composites with planar interfaces [2, 21].

1.4 Effective Physical Properties of Piezo-Active Composites

1.4.1 Piezoelectric Medium and Its Characteristics

An electric polarisation in quartz single crystals subjected to a mechanical load was first reported by brothers P. Curie and J. Curie in 1880 [12–14]. The effect that links a mechanical action (stress or strain) with an electric response of a dielectric sample (electric field, displacement or polarisation) is called *the piezoelectric effect* or, more exactly, the *direct piezoelectric effect*. At a later date in the 1880s, the *converse piezoelectric effect* was first experimentally observed in single crystals where an

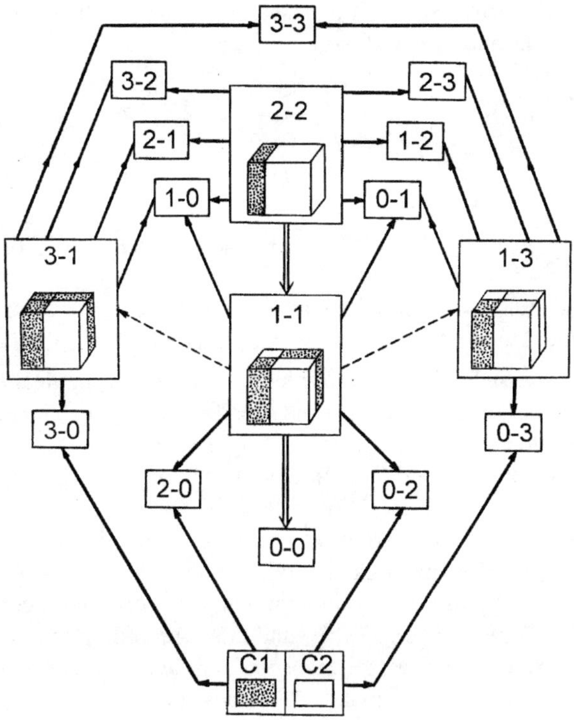

Fig. 1.1 Links between the α–β connectivity patterns. C1 and C2 denote components 1 and 2, respectively. Double lines connecting the 2–2 and 1–1 connectivity patterns as well as the 1–1 and 0–0 connectivity patterns denote a multiplication of structural elements at transformations 2–2 → 1–1 and 1–1 → 0–0. Dashed lines connecting the 1–1 and 1–3 connectivity patterns as well as the 1–1 and 3–1 connectivity patterns denote a simplification of the 1–1 structure at transformations 1–1 → 3–1 and 1–1 → 1–3 (reprinted from Topolov and Glushanin [21], with permission from IOP Publishing)

external electric field generated a mechanical response, i.e. a stress or strain of the sample [13], similar to the electrostriction of dielectrics [12].

The piezoelectric effect reflects a linear relationship between electric and mechanical variables and originates from the displacement of ions of a dielectric single crystal under an applied electric field [12, 13]. It is assumed that this single crystal belongs to one of the 20 symmetry classes listed in Sect. 1.1. Despite the linear character, the piezoelectric response of a single crystal is often intricate owing to various interconnections between the piezoelectric and other properties such as the elastic, dielectric and thermal properties [11–14].

A general consideration of the piezoelectric effect in single crystals is undertaken using thermodynamic functions, such as Helmholz free energy, Gibbs free energy, elastic Gibbs energy and electric Gibbs energy [12, 13]. Each of the aforementioned functions has at least three arguments that characterise the mechanical, electric and thermal states of the single crystal. The first argument can be either mechanical stress σ_{kl} or mechanical strain ξ_{jr}, the second argument can be electric field $\boldsymbol{E}$ or electric displacement (electric flux density) $\boldsymbol{D}$, and the third argument can be temperature T or entropy S. In addition, it is possible to develop additional arguments in terms of magnetic field $\boldsymbol{H}$ or magnetic induction $\boldsymbol{B}$ [3]. Based on knowledge of the thermodynamic functions and relations between the arguments of the mechanical, electric and thermal states, it is possible to describe the response – action links in terms of ξ_{jr}, E_i and T as follows:

$$\sigma_{kl} = c^E_{kljr}\xi_{jr} - e_{ikl}E_i + \beta_{kl}\Delta T \quad (1.1)$$

$$D_i = e_{ikl}\xi_{kl} + \varepsilon^{\xi}_{if}E_f + p_i\Delta T \quad (1.2)$$

$$\Delta S = \beta_{jr}\xi_{jr} + p_iE_i + (\rho/T_0)c\,\Delta T \quad (1.3)$$

In (1.1)–(1.3), the constants written in the tensor form are elastic moduli at constant electric field (c^E_{kljr}), piezoelectric coefficients (e_{ikl}), coefficients of thermal stress (β_{kl}), dielectric permittivities at constant mechanical strain (ε^{ξ}_{if}), pyroelectric coefficients (p_i), density (ρ) and specific heat (c). The increments ΔT and ΔS from (1.1) to (1.3) denote differences $T - T_0$ and $S - S_0$, respectively, where T_0 is initial temperature, and S_0 is initial entropy of the single crystal. It is assumed that linear relations in (1.1)–(1.3) hold in the case of relatively weak external fields. Equations (1.1)–(1.3) are used for the determination of a set of isothermal constants for a piezoelectric single crystal. An analogous set of adiabatic constants can also be derived using the thermodynamic functions and a combination of three arguments including ΔS (e.g. ξ_{jr}, E_i and ΔS). According to experimental data, the difference between the related isothermic and adiabatic constants of the piezoelectric single crystals is approximately 1% or less [12].

The effect of thermal fields on the elastic and electric responses of piezoelectrics is often neglected, and the description of the piezoelectric effect is carried out on the basis of (1.1) and (1.2) at $\Delta T = 0$. Then (1.1) and (1.2) are written in the matrix form [12, 13] as follows:

$$\sigma_p = c^E_{pq}\xi_q - e_{fp}E_f \quad (1.4)$$

$$D_k = e_{kl}\xi_l + \varepsilon^{\xi}_{kr}E_r \quad (1.5)$$

Equations (1.4) and (1.5) represent the first pair of piezoelectric equations linking the variables of mechanical strain and electric field (ξ, E). The first term in the right part of (1.5) is represented as $P_k = e_{kl}\xi_l$. This describes the piezoelectric polarisation produced by an external strain as a result of the direct piezoelectric effect. The converse piezoelectric effect is described by the $e_{fp}\,E_f$ term in (1.4). As follows from the thermodynamic analysis of the interrelations between the electric and elastic fields [12, 13], the converse and direct piezoelectric effects can be described by the three following pairs of equations:

$$\xi_p = s^E_{pq}\sigma_q + d_{fp}E_f \quad (1.6)$$

$$D_k = d_{kl}\sigma_l + \varepsilon^{\sigma}_{kr}E_r \quad (1.7)$$

in variables (σ, E),

$$\xi_p = s_{pq}^D \sigma_q + g_{fp} D_f \tag{1.8}$$

$$E_k = -g_{kl} \sigma_l + \beta_{kr}^\sigma D_r \tag{1.9}$$

in variables (σ, D) and

$$\sigma_p = c_{pq}^D \xi_q - h_{fp} D_f \tag{1.10}$$

$$E_k = -h_{kl} \xi_l + \beta_{kr}^\xi D_r \tag{1.11}$$

in variables (ξ, D). Superscripts σ and D denote measurement conditions at σ = const and D = const, respectively, constants β_{kr}^σ and β_{kr}^ξ are determined using equalities $\beta_{kr}^\sigma \varepsilon_{rv}^\sigma = \delta_{kv}$ and $\beta_{kr}^\xi \varepsilon_{rv}^\xi = \delta_{kv}$, and elastic moduli c_{pq}^E or c_{pq}^D and elastic compliances s_{pq}^E or s_{pq}^D are related by conditions $c_{pq}^E s_{qr}^E = \delta_{pr}$ and $c_{pq}^D s_{qr}^D = \delta_{pr}$, where δ_{kv} is the Kronecker symbol.

In (1.4)–(1.11), we see four types of the piezoelectric coefficients, namely e_{kl}, d_{kl}, g_{kl} and h_{kl}. The piezoelectric coefficient e_{kl} (stress coefficient) relates the electric field to the mechanical stress and the mechanical strain to the electric displacement; see (1.4) and (1.5). The piezoelectric coefficient d_{kl} (charge coefficient) characterises the relation between the electric field and the mechanical strain and the relation between the mechanical stress and the electric displacement; see (1.6) and (1.7). The piezoelectric coefficient g_{kl} (voltage coefficient) is introduced to relate the electric displacement to the mechanical strain and the mechanical stress to the electric field; see (1.8) and (1.9). The piezoelectric coefficient h_{kl} (strain coefficient) links the electric displacement to the mechanical stress and the mechanical strain to the electric field; see (1.10) and (1.11). As seen from (1.4) to (1.11), each piezoelectric coefficient characterises the relationship between components of two fields that are described by first- and second-rank tensors, and all the piezoelectric coefficients are defined to be components of the third-rank tensors, but written in the matrix form (i.e. with two subscripts) [12–14] only. These individual piezoelectric coefficients can be used to aid in the selection of materials for specific piezotechnical applications such as sensors, actuators and acoustic antennae [2, 7–10, 12, 13, 16, 19, 20].

The links between the piezoelectric coefficients follow from the thermodynamic description [12–14] and are given by

$$d_{fp} = \varepsilon_{fk}^\sigma g_{kp} = e_{fq} s_{qp}^E \tag{1.12}$$

$$e_{fp} = \varepsilon_{fk}^\xi h_{kp} = d_{fq} c_{qp}^E \tag{1.13}$$

$$g_{fp} = \beta_{fk}^\sigma d_{kp} = h_{fq} s_{qp}^D \tag{1.14}$$

$$h_{fp} = \beta^{\xi}_{fk}\, e_{kp} = g_{fq}\, c^{D}_{qp} \tag{1.15}$$

We note that the elastic properties (s^{E}_{qp}, c^{E}_{qp}, s^{D}_{qp} and c^{D}_{qp}) and dielectric properties ($\varepsilon^{\sigma}_{fk}$, ε^{ξ}_{fk}, β^{σ}_{fk} and β^{ξ}_{fk}) from (1.12) to (1.15) are present in (1.4)–(1.11). Due to the electromechanical coupling, the piezoelectric coefficients from (1.12) to (1.15) are involved in relations [12–14] between dielectric or elastic constants measured on different conditions:

$$\varepsilon^{\sigma}_{kr} - \varepsilon^{\xi}_{kr} = d_{kf} e_{rf} \tag{1.16}$$

$$\beta^{\xi}_{kr} - \beta^{\sigma}_{kr} = g_{kf} e_{rf} \tag{1.17}$$

$$c^{D}_{pq} - c^{E}_{pq} = e_{fp} h_{fq} \tag{1.18}$$

$$s^{D}_{pq} - s^{E}_{pq} = d_{fp} g_{fq} \tag{1.19}$$

The links between the elastic, piezoelectric and dielectric constants of a piezoelectric medium [see (1.12)–(1.19)] enable us to term these constants as *electromechanical constants*. These constants characterise *electromechanical properties* of piezo-active materials, such as piezoelectric or ferroelectric single crystals, poled FCs, thin ferroelectric films and piezo-active composites.

Examples of the full sets of electromechanical constants s^{E}_{pq}, d_{ij} and $\varepsilon^{\sigma}_{kk}$ of the widespread perovskite-type FCs in the poled state are given in Table 1.2. We add that these constants are involved in (1.6) and (1.7) which are suitable to describe any piezoelectric medium. In the literature, one can find the full sets of electromechanical constants c^{E}_{pq}, e_{ij} and ε^{ξ}_{kk} of the perovskite-type FCs; see, for instance, monographs [2, 3]. Equations (1.4) and (1.5) are written in terms of c^{E}_{pq}, e_{ij} and ε^{ξ}_{kk} and also used to characterise the electromechanical properties of the piezoelectric medium. We remind the reader that matrices of the aforementioned electromechanical constants are written by taking into account symmetry of piezoelectric materials (single crystals, FCs, composites, etc.) [12–14]. In Appendix C, we show examples of the matrices and give a few references on monographs that contain the full sets of electromechanical constants of the poled FCs.

In Table 1.3, we show the piezoelectric coefficients e_{ij}, g_{ij} and h_{ij} of the widespread perovskite-type FCs, and these piezoelectric coefficients have been calculated by taking into account (1.12)–(1.19). The FCs listed in Table 1.2 are characterised by the piezoelectric coefficients that obey conditions $d_{33} > 0$, $d_{31} < 0$, $d_{15} > 0$, $g_{33} > 0$, $g_{31} < 0$ and $g_{15} > 0$. The piezoelectric coefficient e_{31} can be either negative or positive (see Table 1.3), and the equality $\mathrm{sgn} e_{31} = \mathrm{sgn} h_{31}$ holds because of the link in (1.13). The sign of the piezoelectric coefficient e_{31} depends on the anisotropy of the piezoelectric coefficients d_{3f} of the FC and on its elastic properties [2, 7] due to valid (1.13).

Table 1.2 Elastic compliances s_{pq}^E (in 10^{-12} Pa^{-1}), piezoelectric coefficients d_{ij} (in pC/N) and dielectric permittivities ε_{kk}^σ of poled perovskite-type FCs at room temperature [8]

FC	s_{11}^E	s_{12}^E	s_{13}^E	s_{33}^E	s_{44}^E	d_{31}	d_{33}	d_{15}	$\frac{\varepsilon_{11}^\sigma}{\varepsilon_0}$	$\frac{\varepsilon_{33}^\sigma}{\varepsilon_0}$
$BaTiO_3$ (I)	8.55	−2.61	−2.85	8.93	23.3	−79	191	270	1623	1900
$BaTiO_3$ (II)	9.1	−2.7	−2.9	9.5	22.8	−78	190	260	1450	1700
ZTS-19	15.1	−5.76	−5.41	17.0	41.7	−126	307	442	1350	1500
PZT-4	12.3	−4.03	−5.35	15.6	39.1	−124	291	496	1440	1280
PZT-5	16.3	−5.67	−7.17	18.7	47.4	−170	373	583	1730	1700
PZT-5H	10.8	−3.28	−3.41	11.6	28.3	−128	315	482	2640	2490
PZT-7A	10.7	−3.22	−4.62	13.9	39.5	−60.2	151	364	843	427
PZ 27	16.9	−6.32	−8.56	22.5	43.9	−174	419	515	1800	1770
PZ 34	7.71	−1.54	−3.99	13.0	17.0	−3.92	59.1	39.2	191	196
PCR-1, hp[a]	12.5	−4.4	−5.8	15.9	38.8	−95	220	420	1130	650
PCR-7, hp	17.2	−7.2	−6.7	17.3	42.4	−280	610	760	2970	3500
PCR-7M, hp	17.5	−6.7	−7.9	19.6	43.8	−350	760	880	3990	5000
PCR-8, hp	12.5	−4.6	−5.2	15.6	35.3	−130	290	410	1380	1400
PCR-8, ct	13.3	−4.8	−4.4	14.5	39.5	−125	280	458	1320	1300
PCR-13, ct	10.4	−3.7	−2.1	11.3	28.5	−65	140	200	870	780
PCR-21, hp	11.8	−4.5	−3.9	12.6	40.8	−109	250	370	1400	1350
PCR-63, hp	9.8	−3.5	−2.7	9.8	24.1	−60	140	166	960	1170
PCR-73, hp	17.9	−6.8	−9.6	23.5	43.7	−380	860	980	4750	6000

(continued)

Table 1.2 (continued)

FC	s_{11}^E	s_{12}^E	s_{13}^E	s_{33}^E	s_{44}^E	d_{31}	d_{33}	d_{15}	$\varepsilon_{11}^\sigma/\varepsilon_0$	$\varepsilon_{33}^\sigma/\varepsilon_0$
$Pb(Zr_{0.54}Ti_{0.46})O_3$	11.6	−3.33	−4.97	14.8	45.0	−60.2	152	440	990	450
$Pb(Zr_{0.52}Ti_{0.48})O_3$	13.8	−4.07	−5.80	17.1	48.2	−93.5	223	494	1180	730
$(Pb_{0.94}Sr_{0.06})(Ti_{0.47}.Zr_{0.53})O_3$	12.3	−4.05	−5.31	15.5	39.0	−123	289	496	1475	1300
Modified $PbTiO_3$ (I)	7.50	−1.51	−1.10	8.00	17.9	−4.40	51.0	53.0	228	177
Modified $PbTiO_3$ (II)	7.7	−1.7	−1.2	8.2	19	−6.8	56	68	240	190
Modified $PbTiO_3$ (III)	7.51	−1.5	−1.1	8.0	18	−5.0	53	54	230	180
$(Pb_{0.9625}La_{0.025}).(Ti_{0.99}Mn_{0.01})O_3$	7.20	−1.42	−1.73	7.62	15.8	−4.15	47.2	53.0	223	168
$(Pb_{0.85}Nd_{0.10})(Ti_{0.99}\ Mn_{0.01})O_3$	6.84	−1.50	−1.83	7.27	16.4	−5.42	56.8	79.5	313	252
$(Pb_{0.855}Nd_{0.11})(Ti_{0.94}.Mn_{0.02}In_{0.04})O_3$	6.77	−1.55	−1.79	7.29	16.0	−6.40	57.9	86.4	318	248
PMN–0.35PT	13.2	−3.96	−6.05	14.7	33.4	−133	270	936	4610	3270

[a]FC samples of the PCR type have been manufactured using either the conventional technology (ct) or hot pressing (hp). PCR is the abbreviation for the group 'piezoelectric ceramics from Rostov-on-Don' (Russia)

Table 1.3 Piezoelectric coefficients e_{ij} (in C/m^2), g_{ij} (in mVm/N) and h_{ij} (in 10^8 V/m) of poled perovskite-type FCs at room temperature [8][a]

FC	e_{31}	e_{33}	e_{15}	g_{31}	g_{33}	g_{15}	h_{31}	h_{33}	h_{15}
FCs with $e_{33} > 0$ and $e_{31} < 0$									
$BaTiO_3$ (I)	−4.38	18.6	11.6	−4.70	11.4	18.8	−3.48	14.8	10.3
$BaTiO_3$ (II)	−4.32	17.4	11.4	−5.18	12.6	20.3	−3.90	15.7	11.6
ZTS-19	−4.80	15.0	10.6	−9.49	23.1	37.0	−6.43	20.1	14.6
PZT-4	−5.26	15.0	12.7	−10.9	25.7	38.9	−9.32	26.7	19.7
PZT-5	−5.26	15.9	12.3	−11.3	24.2	38.1	−7.18	21.7	15.1
PZT-5H	−6.42	23.4	17.0	−5.81	14.3	27.6	−4.93	17.9	11.2
PZT-7A	−2.27	9.35	9.22	−15.9	49.0	48.8	−10.8	44.7	22.4
PZ 27	−3.59	15.9	11.7	−11.1	26.7	32.3	−4.63	20.5	11.9
PCR-1, hp[b]	−3.81	11.1	10.8	−16.5	38.2	42.0	−14.7	42.6	19.8
PCR-7, hp	−9.10	28.2	17.9	−9.04	19.7	28.9	−10.5	37.5	14.2
PCR-7 M, hp	−9.85	30.8	20.1	−7.91	17.2	24.9	−7.08	28.1	11.4
PCR-8, hp	−7.52	13.6	11.6	−10.5	23.4	33.6	−11.6	20.9	15.6
PCR-8, ct	−6.87	15.1	11.6	−10.9	24.2	39.2	−12.4	27.3	18.2
PCR-13, ct	−6.59	9.94	7.02	−9.42	20.3	26.0	−14.1	21.4	11.1
PCR-21, hp	−6.47	15.8	9.07	−9.12	20.9	29.9	−9.84	24.1	10.0
PCR-63, hp	−4.45	11.8	6.89	−5.79	13.5	19.5	−5.45	14.5	9.37
PCR-73, hp	−8.81	29.4	22.4	−7.16	16.2	23.3	−4.17	13.9	11.2

(continued)

Table 1.3 (continued)

FC	e_{31}	e_{33}	e_{15}	g_{31}	g_{33}	g_{15}	h_{31}	h_{33}	h_{15}
$Pb(Zr_{0.54}Ti_{0.46})O_3$	−1.86	9.02	9.78	−15.1	38.2	50.2	−7.78	37.8	21.9
$Pb(Zr_{0.52}Ti_{0.48})O_3$	−3.08	11.0	10.2	−14.5	34.5	47.3	−8.95	31.8	19.1
$(Pb_{0.94}Sr_{0.06}).(Ti_{0.47}.Zr_{0.53})O_3$	−5.20	15.1	12.7	−10.7	25.1	38.0	−8.87	25.7	18.9
PMN–0.35PT	−5.14	14.1	28.0	−4.60	9.33	22.9	−2.16	5.95	19.2
FCs with $e_{3j} > 0$									
PZ 34	3.82	6.89	2.31	−2.26	34.1	23.2	28.2	50.8	14.4
Modified $PbTiO_3$ (I)	0.459	6.50	2.96	−2.81	32.6	26.3	3.71	52.5	15.9
Modified $PbTiO_3$ (II)	0.247	6.90	3.58	−4.04	33.3	32.0	1.90	53.2	19.0
Modified $PbTiO_3$ (III)	0.401	6.74	3.00	−3.14	33.3	26.5	3.23	54.4	16.0
$(Pb_{0.9625}La_{0.025})\cdot(Ti_{0.99}Mn_{0.01})O_3$	1.31	6.79	3.35	−2.79	31.7	26.9	11.2	57.7	18.7
$(Pb_{0.85}Nd_{0.10})(Ti_{0.99}.Mn_{0.01})O_3$	2.01	8.82	4.85	−2.43	25.5	28.7	11.5	50.4	20.3
$(Pb_{0,855}Nd_{0,11})(Ti_{0,94}.Mn_{0,02}In_{0,04})O_3$	1.80	8.83	5.40	−2.92	26.4	30.7	10.6	51.7	23.0

[a]Based on experimental data from Table 1.2

1.4.2 Methods for Evaluation of Effective Properties

A formulation of a problem on predicting the effective electromechanical properties in a piezo-active composite is given in monograph [3]. If the averaged components of the electric field $\langle E_f \rangle$ and the mechanical strain $\langle \xi_q \rangle$ in a macroscopic region of the composite sample are independent of co-ordinates x_j ($j = 1$, 2 and 3), then (1.4) and (1.5) are written for the piezo-active composite (heterogeneous medium) as

$$\langle \sigma_p \rangle = c_{pq}^{*E} \langle \xi_q \rangle - e_{fp}^{*} \langle E_f \rangle \tag{1.20}$$

$$\langle D_k \rangle = e_{kl}^{*} \langle \xi_l \rangle + \varepsilon_{kr}^{*\xi} \langle E_r \rangle \tag{1.21}$$

In (1.20) and (1.21), elastic moduli c_{pq}^{*E}, piezoelectric coefficients e_{fp}^{*} and dielectric permittivities $\varepsilon_{kr}^{*\xi}$ constitute the full set of effective electromechanical constants that are determined [3] by taking into account equations of electric and mechanical equilibrium and boundary conditions at the surface of the macroscopic region in the heterogeneous medium. Hereafter we use asterisk (*) to denote the effective (average) properties of the composite. The determination of the effective electromechanical properties of the piezo-active composite is concerned with an averaging of a series of vector and tensor components of the electric and mechanical fields, for example, the mechanical strain ξ_q, electric field E_f, mechanical stress σ_p and electric displacement D_k from (1.4) to (1.5). This averaging is performed on volume fractions of the composite components ($n = 1, 2, \ldots$), for which the full sets of electromechanical constants (e.g. $c_{pq}^{(n),E}$, $e_{fp}^{(n)}$ and $\varepsilon_{kr}^{(n),\xi}$) are known.

Based on the detailed analysis of results concerning piezo-active composites and their effective properties, Khoroshun et al. [3] emphasised two main scientific directions in modern mechanics of piezo-active composites. The first direction represents the mechanics of the composites having regular (periodic) structures. The second direction is related to the mechanics of stochastically heterogeneous materials, including the piezo-active composites with arbitrary connectivity, shape and distribution of structural elements.

The first direction deals with periodic solutions of equations of mechanics of micro-inhomogeneous media in which the regular distribution of the structural elements is present. As a rule, such solutions are applicable to the matrix composites with inclusions having the appropriate shape, for instance, spheroids. The accompanying method, termed *the method for regularisation of structure* [3], is often applied to the piezo-active composites with a certain disorder in the shape and distribution of the structural elements. In this method, the real composite sample is substituted by a model material with regularly distributed structural elements having the appropriate shape. In such a model material, a periodic representative volume element is chosen and analysed using the appropriate micromechanical methods, including the widespread *finite element method* (FEM) [7].

The second direction implies using random functions [3] to describe the effective properties of the piezo-active composite with a stochastic structure. These properties are regarded as randomly varied ones over the composite sample, and the homogenisation can be performed with regard to mathematical features of the random functions. This approach is suitable for the prediction of the effective properties without essential restrictions of the shape, distribution of inclusions and the connectivity of the composite as a whole.

The matrix method is based on the matrix approach [22] applied to the 2–2 composite system [2, 7, 8] with a regular distribution of layers. Many publications on 0–3 composites are concerned with a piezoelectric ellipsoidal inclusion problem [2, 3, 7, 8, 23, 24]. Coupled electric and mechanical fields in heterogeneous materials are analysed by means of the equivalent inclusion method [25, 26] for the description of the effective properties in the piezo-active composites, poled FCs, porous materials, etc. Further averaging of the electromechanical properties is often carried out according to a self-consistent scheme. The main variants of this scheme are concerned with the *effective medium method* [7, 27, 28] and the *effective field method* (also termed *the mean field method*) [7, 29, 30]. In some cases, especially when studying a three-component composite, a composite structure is often regarded as '*a composite in composite*', and for an evaluation of the effective properties, two methods or a few subsequent procedures of averaging can be applied [2, 7, 8, 31].

The determination of the effective properties of the piezo-active composite with arbitrary connectivity is also carried out by means of the FEM [7, 31, 32]. This method employs numerical modelling techniques that take into account boundary conditions for electric and mechanical fields in a composite system. The model of the composite is based on the representative unit cell being the periodic representative volume element, and this element is assumed to be characteristic of the whole composite structure. Periodic boundary conditions are enforced on the boundary of the representative unit cell. The latter is discretised using tetrahedral or hexahedral elements in 3D problems, or using triangular or quadrilateral elements in 2D problems. The unknown displacement and electric potential fields are interpolated using suitable shape functions: usually, linear or quadratic Lagrangian shape functions are adopted. Computations for diverse average strain and electric fields imposed to the representative unit cell are performed. Direct or iterative solvers can be used to solve the resulting linear system of equations. After solving the electroelastic equilibrium problem, the effective electromechanical constants of the composite are computed column-wise, by averaging the resulting local stress and electric-displacement fields over the representative unit cell.

Theoretical models used to predict the effective electromechanical properties of the composites based on ferroelectrics can be divided into three groups [33] as follows. In the group 1, no interactions between components are assumed, and the composite system is approximated by dielectric spheres dispersed in a continuous matrix. In the group 2, interactions between a filler with neighbouring particles under an electric field applied to the composite sample are assumed. In the group 3, the shape and orientation of the inclusions in the composite is taken into account.

An example of the effective piezoelectric properties of the 1–3 and 0–3 FC/polymer composites (Fig. 1.2) is concerned with the longitudinal piezoelectric effect. This means that an external field (either mechanical stress or electric field)

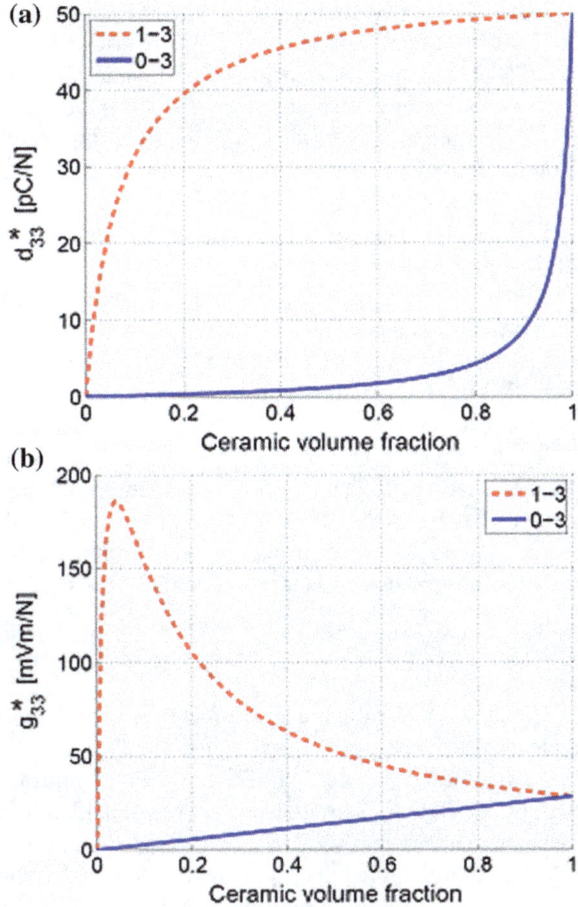

Fig. 1.2 Longitudinal piezoelectric coefficients d_{33}^* **a** and g_{33}^* **b** of 0–3 and 1–3 $PbTiO_3$-type FC/epoxy composites at various volume fractions of FC [33]

acts along the OX_3 direction, that is the poling direction for the composite sample, and the response of this sample is detected on the OX_3 direction. We remind the reader that for the piezoelectric coefficients d_{33}^* and g_{33}^*, (1.6)–(1.9) can be used to express the links between the external field and response of the composite sample.

The effective properties of the composite can be evaluated by using different methods and approaches; see, for instance, [2, 3, 7, 8, 10, 22, 33, 34]. In Fig. 1.2, we show the results of calculations performed within the framework of the mixed connectivity model (MCM) [33]. As follows from Fig. 1.2a, large values of the piezoelectric coefficient d_{33}^* of the 0–3 composite are achieved at very large volume fractions of FC. In contrast to this, the piezoelectric coefficient d_{33}^* of the 1–3 composite approaches $d_{33}^{(1)}$ of the FC component at the volume fractions over 0.3. Figure 1.2b suggests that the piezoelectric coefficient g_{33}^* of the 0–3 composite remains relatively small in the whole volume-fraction range in comparison to g_{33}^* of the 1–3 composite. The reason of the different volume-fraction dependences shown in Fig. 1.2 is mainly accounted for the specifics of the composite microgeometry. In the 0–3 composite, the FC component (i.e. the main piezoelectric component) is distributed in the form of isolated inclusions, and in the 1–3 composite the same FC component is distributed continuously along the poling axis OX_3 (see Table 1.1). Moreover, due to the lower resistivity and larger permittivity of the FC component, the electric flux concentrates more effectively on the polymer component (i.e. the polymer matrix in the studied composites) and does not affect the FC particles efficiently. At high volume fractions of FC, the thickness of the polymer layer separating the FC particles in the composite sample becomes smaller. Such a lower polymer layer thickness facilitates the poling of the composite, as a smaller fraction of the applied external electric field is imposed on the polymer matrix rather than on the poling of the FC particles. The 1–3 composite exhibits in general a higher piezoelectric performance in comparison to its 0–3 analog [2] due to the continuous distribution of the FC component along the poling direction.

The relatively sharp max g_{33}^* in Fig. 1.2b is caused by a rapid increase of the piezoelectric coefficient d_{33}^* at small volume fractions of FC (see Fig. 1.2a) and by a slow increase of the dielectric permittivity $\varepsilon_{33}^{*\sigma}$ of the composite. We remind the reader that the piezoelectric coefficient g_{33}^* of the FC/polymer composite poled along the OX_3 axis (∞mm symmetry) is defined in accordance with (1.12) as

$$g_{33}^* = d_{33}^* / \varepsilon_{33}^{*\sigma}. \tag{1.22}$$

Equation (1.22) illustrates an effect of a combination of the piezoelectric (d_{33}^*) and dielectric ($\varepsilon_{33}^{*\sigma}$) properties in the piezo-active composite [2, 5]. It should be added that examples of max g_{33}^* in FC/polymer composites with different connectivity patterns (1–3, 2–2, 3–3, 1–1, etc.) were considered in many publications; see, for instance, monographs [2, 7, 8]. The volume-fraction dependences shown in Fig. 1.2 are consistent with results [2] on the 0–3 and 1–3 composites based on $PbTiO_3$-type FCs. Large values of the piezoelectric coefficient $g_{33}^* > 100$ mVm/N (see Fig. 1.2b) that are achieved at the relatively small volume fractions of the FC component in the 1–3 composite are of interest in piezoelectric sensor applications [9, 14].

1.4.3 Electromechanical Coupling Factors and Figures of Merit

Full sets of electromechanical constants from (1.12) to (1.15) are of value to estimate an effectiveness of a conversion of electric energy into mechanical energy and vice versa. This effectiveness depends [12, 13] on differences between dielectric permittivities from (1.16) or differences between elastic compliances from (1.19), and these differences are caused by the piezoelectric effect. In a general form, the effectiveness of the energy conversion in the piezoelectric medium is described by an electromechanical coupling factor (ECF) [12–14] $k = w_{piezo}/\sqrt{w_{el} w_{mech}}$ wherein the volume density of piezoelectric (or mutual) energy $w_{\text{piezo}} = \sigma_i\, d_{mi} E_m/2$, the volume density of electric energy $w_{\text{el}} = E_l \varepsilon^{\sigma}_{lr} E_r/2$, and the volume density of mechanical energy $w_{\text{mech}} = \sigma_p s^E_{pq} \sigma_q/2$ are involved. The k^2 value characterises a ratio of stored mechanical energy to electric energy input (when a portion of electric energy is applied to the piezoelectric sample) or a ratio of stored electric energy to mechanical energy input (when a portion of mechanical energy is applied to the piezoelectric sample) [12–14]. The absolute ECF value depends on the vibration mode, measurement methodology, sample shape and the electromechanical constants of the piezoelectric medium. As a rule, two kinds of ECFs are considered as follows: the static ECF determined from equations of the state of the piezoelectric medium and the dynamic (effective) ECF related to the individual piezoelectric element [12, 13].

A poled FC element with a remanent polarisation vector $\boldsymbol{P_r} || OX_3$ (∞mm symmetry) is characterised by three independent piezoelectric coefficients [12–14], namely d_{31}, d_{33} and d_{15}. The ECFs are expressed in terms of d_{ij} as follows:

$$k_{33} = d_{33}/\left(\varepsilon^{\sigma}_{33}\, s^E_{33}\right)^{1/2} \tag{1.23}$$

(ECF at the longitudinal oscillation mode, or simply, longitudinal ECF),

$$k_{31} = d_{31}/\left(\varepsilon^{\sigma}_{33}\, s^E_{11}\right)^{1/2} \tag{1.24}$$

(ECF at the transverse oscillation mode) and

$$k_{15} = d_{15}/\left(\varepsilon^{\sigma}_{11}\, s^E_{55}\right)^{1/2} \tag{1.25}$$

(ECF at the shear oscillation mode). Since an equality $s^E_{55} = s^E_{44}$ holds [12, 13] in the poled FC, (1.25) is often written as

$$k_{15} = d_{15}/\left(\varepsilon^{\sigma}_{11}\, s^E_{44}\right)^{1/2} \tag{1.26}$$

It should be added that for the poled FC, $k_{15} = k_{24}$ because conditions $d_{15} = d_{24}$, $s^E_{55} = s^E_{44}$ and $\varepsilon^{\sigma}_{11} = \varepsilon^{\sigma}_{22}$ hold. By the similar way we state that $k_{31} = k_{32}$ for the poled FC.

The ECF of any piezoelectric medium is written in the general form as

$$k_{ij} = d_{ij}/\left(\varepsilon^{\sigma}_{ii}\, s^E_{ij}\right)^{1/2} \tag{1.27}$$

Along with the ECFs from (1.23) to (1.26), the following ECFs are often measured on poled FC and piezo-active composite samples with ∞mm symmetry [2, 3, 8, 12–14]:

$$k_{ij} = d_{ij}/\left(c^D_{33}\, \varepsilon^{\xi}_{33}\right)^{1/2} \tag{1.28}$$

(ECF at the thickness oscillation mode, or simply, thickness ECF),

$$k_p = k_{31}[2/(1-\sigma_E)]^{1/2} \quad (1.29)$$

(ECF at the planar oscillation mode, or simply, planar ECF) and

$$k_h = d_h/\left(\varepsilon_{33}^{\sigma}\, s_h^E\right)^{1/2} \quad (1.30)$$

(hydrostatic ECF). In (1.29), σ_E is the Poisson ratio at electric field $E = \text{const}$, and in (1.27) d_h and s_h^E are the hydrostatic piezoelectric coefficient and hydrostatic elastic compliance at $E = \text{const}$, respectively. Data on the ECFs are useful in the selection of piezoelectric materials for active elements of electromechanical transducers, hydrophones and piezoelectric energy harvesters [3, 8–10, 14–16].

There are three major stages [8, 35] concerned with piezoelectric energy harvesting in electromechanical systems. These stages are characterised as follows:

(i) mechanical–mechanical energy transfer, including the mechanical stability of the piezoelectric transducer under a large mechanical stress, and mechanical impedance matching,
(ii) mechanical–electric energy transduction due to electromechanical coupling in a piezoelectric element and
(iii) electric–electric energy transfer, including electric impedance matching, for instance, in the presence of a DC/DC converter to accumulate the energy into an electric rechargeable battery.

Mechanical (acoustic) impedance matching is one of the important factors to be taken into account at the first stage [35]. The mechanical impedance of a material is given by $Z = (\rho c_{ij})^{1/2}$, where ρ is its density, and c_{ij} is its elastic modulus. At the second stage, a portion of mechanical energy that reaches the piezoelectric transducer is converted into electric energy due to electromechanical coupling. Hereby a voltage induced in the piezoelectric transducer is written in the general form as

$$V = g_{ij} Ft/A \quad (1.31)$$

In (1.31), g_{ij} is the piezoelectric coefficient, F is the applied force, t is the thickness of the piezoelectric element, and A is the area of the surface on which the force acts. Taking into account the mechanical loading and piezoelectric effect, an output electric power of the transducer is written as

$$P = \left(CV^2/2\right) f \quad (1.32)$$

In (1.32), C is the electric capacitance of the piezoelectric element, and f is frequency of its vibration. We note that the capacitance C is proportional to dielectric permittivity $\varepsilon_{jj}^{\sigma}$ for a flat large electric capacitor, and f is chosen in a low-frequency range far from a resonance frequency of the piezoelectric element [35].

When the mechanical loading is applied along the poling axis OX_3 of the FC or composite sample [13–15] with electrodes that are perpendicular to OX_3, the output electric power is $P_{out} \sim d_{33}g_{33}$. Hereafter

$$(Q_{33})^2 = d_{33}g_{33} \tag{1.33}$$

is the squared figure of merit [2, 7–10, 35–37] concerned with the longitudinal piezoelectric effect and oscillation mode [see subscripts '33' in (1.33)]. Taking into account (1.32) and (1.33), we represent energy generated at the applied force F as $W_{gen} = (Q_{33})^2 F^2 t/(2lw)$, where l is the length, and w is the width of the piezoelectric element in the form of the rectangular parallelepiped.

Along with $(Q_{33})^2$ related to the longitudinal piezoelectric effect, we introduce squared figures of merit

$$(Q_{31})^2 = d_{31}g_{31} \text{ and } (Q_{32})^2 = d_{32}g_{32} \tag{1.34}$$

which are related to the transverse piezoelectric effect and oscillation 31 and 32 modes, respectively. At hydrostatic loading, the squared figure of merit

$$(Q_h)^2 = d_h g_h \tag{1.35}$$

is of importance. We note that $(Q_{31})^2$, $(Q_{32})^2$ and $(Q_h)^2$ from (1.34) to (1.35) are analogs of $(Q_{33})^2$ from (1.33), and in every case we see the multiplication of the piezoelectric coefficients d (piezoelectric activity) and g (piezoelectric sensitivity). The squared figures of merit (1.33)–(1.35) belong to the important parameters concerned with piezoelectric energy harvesting [8, 37] and are used to characterise the sensor signal-to-noise ratio of the piezo-active composite and its sensitivity [2, 9, 16, 36] in a specific frequency range

1.5 Conclusion

The present chapter has been devoted to the introduction of piezo-active composites and description of their physical properties and related characteristics. There are various criteria of the classification of the composites; however, their microgeometry and connectivity play the leading role. Knowledge of the connectivity pattern enables us to carry out a prediction of the effective physical properties and related parameters in terms of various methods that take into account the constitutive relations in the piezoelectric medium [see (1.4)–(1.11)] and specifics of the microgeometry, boundary conditions for electric and mechanical fields, etc. [2, 3, 7, 8, 36].

The piezo-active composites play the important role among modern active dielectrics due to the considerable electromechanical coupling, piezoelectric activity, sensitivity and anisotropy, figures of merit, etc. This is achieved in the presence of highly effective ferroelectric components such as poled FCs and domain-engineered relaxor-ferroelectric single crystals [2, 7–10, 37]. The properties of the composites are regarded as effective properties in accordance with features of the microstructure, domain structure, arrangement of components and connectivity. The piezo-active composites based on ferroelectrics are the final link in the hierarchy-of-properties chains of 'single-domain ferroelectric single crystal → polydomain ferroelectric single crystal → FC → composite based on FC' and 'single-domain ferroelectric single crystal → polydomain ferroelectric single crystal → heterophase ferroelectric single crystal → composite based on ferroelectric FC'. Such a status makes the piezo-active composites a unique range of modern functional materials whose effective physical properties can be tailored in wide ranges due to a range of factors including physical, chemical, microgeometric and technological. The remarkable piezoelectric properties and related parameters of the composites based on ferroelectrics stimulate a creation of novel highly effective materials and are to be taken into account for potential piezotechnical and energy-harvesting applications.

References

1. R.E. Newnham, Molecular mechanisms in smart materials. Mater. Res. Soc. Bull. **22**(5), 20–34 (1997)
2. V.Yu. Topolov, C.R. Bowen, *Electromechanical Properties in Composites Based on Ferroelectrics* (Springer, London, 2009)
3. L.P. Khoroshun, B.P. Maslov, P.V. Leshchenko, *Prediction of Effective Properties of Piezo-Active Composite Materials* (Naukova Dumka, Kiev, 1989) (in Russian)
4. R.M. Chistensen, *Mechanics of Composite Materials* (Wiley, New York, 1979)
5. R.E. Newnham, Nonmechanical properties of composites, in *Concise Encyclopedia of Composite Materials*, ed. by A. Kelly, R.W. Cahn, M.B. Bever (Elsevier, Oxford, 1994), pp. 214–220
6. R.E. Newnham, D.P. Skinner, L.E. Cross, Connectivity and piezoelectric-pyroelectric composites. Mater. Res. Bull. **13**, 525–536 (1978)
7. V.Yu. Topolov, P. Bisegna, C.R. Bowen, *Piezo-active composites. Orientation Effects and Anisotropy Factors* (Springer, Berlin Heidelberg, 2014)
8. C.R. Bowen, V.Yu. Topolov, H.A. Kim, *Modern Piezoelectric Energy-Harvesting Materials* (Springer International Publishing Switzerland, 2016)
9. E.K. Akdogan, M. Allahverdi, A. Safari, Piezoelectric composites for sensor and actuator applications. IEEE Trans. Ultrason. Ferroelectr. Freq. Control **52**, 746–775 (2005)
10. C.R. Bowen, V.Yu. Topolov, A.N. Isaeva, P. Bisegna, Advanced composites based on relaxor-ferroelectric single crystals: from electromechanical coupling to energy-harvesting applications. CrystEngComm **18**, 5986–6001 (2016)
11. M. Lines, A. Glass, *Principles and Application of Ferroelectrics and Related Materials* (Clarendon Press, Oxford, 1977)
12. I.S. Zheludev, *Physics of Crystalline Dielectrics. V. 2: Electrical Properties* (Plenum, New York, 1971)
13. T. Ikeda, *Fundamentals of Piezoelectricity* (Oxford University Press, Oxford New York Toronto, 1990)
14. J. Tichý, J. Erhart, E. Kittinger, J. Přivratská, *Fundamentals of Piezoelectric Sensories. Mechanical, Dielectric, and Thermodynamical Properties of Piezoelectric Materials* (Springer, Berlin, Heidelberg, 2010)

15. B. Jaffe, W.R. Cook, H. Jaffe, *Piezoelectric Ceramics* (Academic Press, London New York, 1971)
16. C.H. Sherman, J.L. Butler, *Transducers and Arrays for Underwater Sound* (Springer, New York, 2007)
17. R. Zhang, B. Jiang, W. Cao, Elastic, piezoelectric, and dielectric properties of multidomain $0.67Pb(Mg_1/3Nb_2/3)O_3$–$0.33PbTiO_3$ single crystals. J. Appl. Phys. **90**, 3471–3475 (2001)
18. R. Zhang, B. Jiang, W. Cao, A. Amin, Complete set of material constants of $0.93Pb(Zn_{1/3}Nb_{2/3})O_3$–$0.07PbTiO_3$ domain engineered single crystal. J. Mater. Sci. Lett. **21**, 1877–1879 (2002)
19. J.E. Smay, B. Tuttle, J. Cesarano III, Robocasting of three-dimensional piezoelectric structures, in *Piezoelectric and Acoustic Materials for Transducer Applications*, ed. by A. Safari, E.K. Akdoğan (Springer, New York, 2008), pp. 305–318
20. F. Wang, C. He, Y. Tang, X. Zhao, H. Luo, Single-crystal $0.7Pb(Mg_{1/3}Nb_{2/3})O_3$–$0.3PbTiO_3$ /epoxy 1–3 piezoelectric composites prepared by the lamination technique. Mater. Chem. Phys. **105**, 273–277 (2007)
21. V.Yu. Topolov, S.V. Glushanin, Evolution of connectivity patterns and links between interfaces and piezoelectric properties of two-component composites. J. Phys. D Appl. Phys. **35**, 2008–2014 (2002)
22. F. Levassort, M. Lethiecq, D. Certon, F. Patat, A matrix method for modeling electroelastic moduli of 0–3 piezo-composites. IEEE Trans. Ultrason. Ferroelectr. Freq. Control **44**, 445–452 (1997)
23. M.L. Dunn, M. Taya, An analysis of piezoelectric composite materials containing ellipsoidal inhomogeneities. Proc. R. Soc. (Lond.), Pt A **443**, 265–287 (1993)
24. J.H. Huang, S. Yu, Electroelastic Eshelby tensors for an ellipsoidal piezoelectric inclusion. Compos. Eng. **4**, 1169–1182 (1994)
25. J. Eshelby, The determination of the elastic field of an ellipsoidal inclusion, and related problems. Proc. R. Soc. (Lond.), Pt A **241**, 376–396 (1957)
26. J. Eshelby, The elastic field outside an ellipsoidal inclusion. Proc. R. Soc. (Lond.), Pt A **252**, 561–569 (1959)
27. C.-W. Nan, Effective-medium theory of piezoelectric composites. J. Appl. Phys. **76**, 1155–1163 (1994)
28. N. Fakri, L. Azrar, L. El Bakkali, Electroelastic behavior modeling of piezoelectric composite materials containing spatially oriented reinforcements. Int. J. Solids Struct. **40**, 361–384 (2003)
29. T. Mori, K. Tanaka, Average stress in matrix and average elastic energy of materials with misfitting inclusions. Acta Metall. **21**, 571–574 (1973)
30. J.H. Huang, W.-S. Kuo, Micromechanics determination of the effective properties of piezoelectric composites containing spatially oriented short fibers. Acta Mater. **44**, 4889–4898 (1996)
31. V.Yu. Topolov, C.R. Bowen, P. Bisegna, S.E. Filippov, The piezoelectric performance and anisotropy factors of modern three-component composites, in *Nano- and Piezoelectric Technologies, Materials and Devices*, ed. by I.A. Parinov (Nova Science Publishers, New York, 2013), pp. 51–78
32. V.Yu. Topolov, P. Bisegna, C.R. Bowen, Analysis of the piezoelectric performance of modern 0–3-type composites based on relaxor-ferroelectric single crystals. Ferroelectrics **413**, 176–191 (2011)
33. H. Khanbareh, Expanding the functionality of piezo-particulate composites. Proefschrift ter verkrijging van der grad van doctor aan de Technische Universiteit Delft (Delft, 2016)
34. Yu.V. Sokolkin, A.A. Pan'kov, *Electroelasticity of Piezo-Composites with Irregular Structures* (Fizmatlit, Moscow, 2003) (in Russian)
35. K. Uchino, T. Ishii, Energy flow analysis in piezoelectric energy harvesting systems. Ferroelectrics **400**, 305–320 (2010)
36. V.Yu. Topolov, C.R. Bowen, P. Bisegna, *Piezo-Active Composites. Microgeometry—Sensitivity Relations*. (Springer International Publishing, Cham, 2018)
37. C.R. Bowen, V.Yu. Topolov, Y. Zhang, A.A. Panich, 1–3-type composites based on ferroelectrics: electromechanical coupling, figures of merit, and piezotechnical energy-harvesting applications. Energy Technology **6**, 813–828 (2018)

Chapter 2
Aspects of Composite Manufacturing

Abstract Methods to manufacture modern composites based on ferroelectrics are described and compared. The overwhelming majority of the well-approved methods are concerned with manufacturing of the ferroelectric ceramic/polymer composites that have been widespread for the past decades. The methods to manufacture the piezo-active composites are based on knowledge of physics, chemistry, materials science and technological principles. In the present work, the main emphasis is placed on applications of dielectrophoresis to form long chains of aligned and closely connected ceramic particles in an electric field, when the matrix polymer is in a low-viscosity state. The methods are suitable for the manufacture piezo-active 0–3, 1–3 and quasi 1–3 ceramic/polymer composites including those for sensor and high-temperature applications. A ceramic-based composite with a foam polymer matrix is also manufactured, and its performance is compared to that of a related 0–3 ceramic/polymer composite.

Methods for the design and manufacture of modern composites based on ferroelectrics are examined and developed [1–6] to obtain highly effective materials with stable and predictable properties [7], to optimise specific properties and related parameters for piezo-technical and other transducer applications. Particular properties of composites are shown to strongly differ from the same properties of the composite components [1, 3, 7] and can be achieved, for instance, in the presence of specific connectivity patterns [4, 8]. This is a result of a combination of properties of the components [7], due to an addition of an extra-component [6], due to an influence of electric-field structuring [9, 10], etc. The overwhelming majority of the methods to manufacture piezo-active composites are related to the FC/polymer composites [1–4], and these methods have been developed since the 1980s. As follows from papers published in the past decades, there are also methods that can be applied to manufacture ferroelectric single crystal/polymer composites [5], ferroelectric single crystal/FC composites [11, 12], FC/cement [13], FC/clay [14], FC/sand [15] and related composites that contain FC and construction materials.

In this chapter, we discuss methods that enable one to manufacture piezo-active composites with predictable and reproducible electromechanical properties.

H. Khanbareh et al., *Piezo-Particulate Composites*, Springer Series in Materials Science 283, https://doi.org/10.1007/978-3-030-19204-4_2

2.1 Methods for Manufacturing

One can divide the methods into the following groups:

(i) methods suitable for manufacturing of the FC/polymer composites,
(ii) methods suitable for manufacturing of ferroelectric single crystal/polymer composites,
(iii) methods suitable for manufacturing of ferroelectric single crystal/FC composites, and
(iv) methods suitable for manufacturing of three-component FC-based composites.

The methods (i)–(iv) can be also divided into the methods that lead to forming the regular composite structures including those concerned with specific connectivity patterns [4, 5] and the methods to form stochastic composite structures, piezoelectric paints and mixtures. In general, the methods (i)–(iv) are based on knowledge of physics, chemistry, materials science and technological principles. Problems related to the piezo-active composites and trends in the development of methods to manufacture these materials with improved properties are highlighted and discussed in a number of review papers; see, for instance [1–4, 2, 16].

The methods based on the computer-aided design (CAD) [3, 4] are useful to manufacture FC/polymer composites with a complicated microgeometry, and complexity of the composite architecture can be higher than other methods developed in the last decades. The use of CAD-based methods stimulate the development of robocasting and rapid prototyping [3, 4, 16–18] to manufacture composite samples with reproducible properties, at specific volume fractions of components, etc. These methods are applied, for instance, to manufacture the 3–β FC/polymer composites [17, 18] with effective parameters that pass maxima and are of value for sensor, hydroacoustic and other applications [3, 4]. The use of CAD opens up new opportunities to manufacture the piezo-active composite samples whose quality is higher than in other widespread methods [1, 3] such as dice and fill, injection moulding, tape lamination and so on. The samples obtained within the framework of the CAD-based methods [2–4] can have structural elements (or fine-scale structure) whose sizes can be smaller than those in samples manufactured by means of the above other methods [1] that have been applied since the 1980s.

Structuring FC particles in the FC/polymer composite by means of dielectrophoresis (DEP) is of value due to an improvement of the piezoelectric properties and some related parameters of this material in comparison with the similar parameters of a piezo-active composite with a random distribution of the FC component. DEP is actively used to form long chains of aligned and closely connected FC particles in an electric field [9, 10, 19], in particular when the matrix polymer is still in its low-viscosity state. The FC/polymer composites manufactured by means of DEP are also termed '*piezo-particulate composites*' and '*quasi 1–3 composites*' [20]. Their improved electromechanical properties [19, 20] can be explained by the higher degree of parallel connectivity of the FC component in comparison with the traditional 0–3 FC/polymer composite [1, 7] wherein systems of isolated FC inclusions do not promote a high piezoelectric activity in a composite sample, even after careful poling in a strong electric field.

A development of the selective laser sintering method enabled Gureev et al. [21] to manufacture novel piezo-active FC-based composites with improved properties. This method is applied to manufacture composites with various connectivity patterns by using powder components (e.g. FC, glass or metal). The main stages of this method are consecutive placing the layers with appointed compositions and sintering by using the laser radiation [21, 22]. After these stages, the composite samples can undergo an additional thermal treatment, be covered with special layers etc., to increase mechanical strength of the samples and improve their specific characteristics.

The ferroelectric single crystal/FC composites from work [11, 12] can be regarded as an alternative to conventional FC/polymer composites. In work [11] components of the piezo-active composite are well-known ferroelectric $PbTiO_3$ single crystal (inclusions) and $(Pb_{1-x}Ca_x)TiO_3$ FC (matrix). As follows from microstructure data obtained by means of the optical microscopy, a distribution of the ferroelectric single-crystal inclusions in the FC matrix is relatively homogeneous, and the isolated character of the distribution of these inclusions suggests that the composite studied in work [11] can be characterised by a 0–3 connectivity pattern.

In terms of the development of methods for manufacturing ferroelectric single crystal/FC composites, a hybrid technology was introduced and reported [12]. This technology consists in a high-frequency heating of FC and ferroelectric single-crystal components and a hot pressing of a composite sample in a specific temperature range. An electromagnetic field at a frequency f in the 10^{10} Hz range is applied for heating. The composite samples manufactured and studied in work [12] are almost free of microcracks and show definite boundaries between the single-crystal inclusions and FC matrix. The composite is characterised by connectivity 0–3, and in the poled state, this composite exhibits a better piezoelectric response in comparison with the poled FC matrix [12].

A modification of the FC-based composites [23–25] is to be taken into account in the context of their piezoelectric properties including sensitivity and anisotropy. For example, piezo-passive oxide inclusions (PbO, MgO, SiO_2, etc.) in a FC matrix of composites [23–25] influence their piezoelectric properties and related parameters to a large extent. Even at volume fractions of the oxide inclusions near 1% and their linear sizes within the 10 nm range, a considerable decrease of the piezoelectric activity of the composite is observed in comparison with its FC component [24].

In Table 2.1, we list the methods that can be applied to manufacture the two-component FC/polymer composites. We show examples for five α–β connectivities that are of interest due to the piezoelectric performance of the related composites [1–8]. It can be seen from Table 2.1 that the rapid prototyping, for instance, can be implemented for any connectivity pattern, and DEP can be applied to manufacture the piezo-active 0–3 and/or 1–3 composites. Many examples of the effective application of rapid prototyping and related CAD-based methods are considered in work [3, 4].

Table 2.1 Methods for manufacturing the α–β FC/polymer composites [1–4, 7]

Connectivity α–β	Methods for manufacturing
2–2	Dice and fill, injection moulding, tape lamination, jet machining, co-extrusion, rapid prototyping
1–3	Rod placement (pick and place), dice and fill, injection moulding, co-extrusion, tape lamination, lost mould, DEP, robocasting, rapid prototyping
0–3	DEP, pressing mixture of ceramic and polymer, colloid processing method, mixing and hot rolling, rapid prototyping
3–3	Lost-wax method and replamine process, burned-out plastic spheres technique, relic procession, distorted reticulated ceramics, robocasting, rapid prototyping
1–1	Laminating the layers, dice and fill, rapid prototyping

2.2 Role of Ceramic and Polymer Components

Knowledge of electromechanical properties of FCs (Table 1.2) and polymers (Table 2.2) opens up ways to select components and manufacture piezo-active composites with predictable and reproducible properties. The role of the specific FC and polymer components that promote an improvement of the composite properties is discussed in a number of papers and monographs [3, 7–9, 26–29]. This role depends on the connectivity index ($\alpha, \beta, \ldots$) of the component, features of the microgeometry on the local composite level and other factors.

In the FC/polymer composites, the following effects [7, 8] related to forming the effective electromechanical properties in these materials are observed:

(i) the effect of the addition of the properties (forming the *sum* properties), and
(ii) the effect of the combination of the properties (forming the *combination* properties).

When forming the sum property Y^* of a two-component composite, we come to the condition

Table 2.2 Elastic compliances s_{ab} (in 10^{-12} Pa^{-1}) and dielectric permittivity ε_{pp} of piezo-passive polymers [26–30] at room temperature

Polymer	s_{11}	s_{12}	$\varepsilon_{pp}/\varepsilon_0$
Araldite	216	−78	4.0
Polyurethane	405	−151	3.5
Elastomer	3300	−1480	5.0
Polyethylene (high-density)	1540	−517	2.3
Auxetic polyethylene	5260	4360	2.3

$$Y^{(1)} < Y^* < Y^{(2)}. \qquad (2.1)$$

In (2.1), $Y^{(n)}$ is the property of the nth component, and $n = 1$ and 2. Y^* is regarded as a sum property and can be described in a simplest case by the formula $\Delta Y^*(m) = Y^{(1)}m + Y^{(2)}(1 - m) + \Delta Y$, where m is the volume fraction of the first component, and ΔY depends on $Y^{(n)}$ and m and characterises deviations of $Y^*(m)$ from the linear dependence [7]. Hereby, a monotonic $Y^*(m)$ dependence is often observed [7, 31, 32]. For the FC/polymer composite, one can achieve a desired value of the effective property Y^* from the range (2.1), and this property is restricted by the similar properties of the FC and polymer components. Validity of (2.1) is confirmed, e.g. for the piezoelectric coefficients d_{ij}^* and dielectric permittivities $\varepsilon_{pp}^{*\sigma}$ of some 1–3 composites [7, 29, 31, 32] that contain PZT-type FCs and piezo-passive polymers.

In a case of the effect of the combination, the properties $Y^{(n)}$ and $Z^{(n)}$ of the components of the composite ($n = 1$ and 2) form the effective properties $Y^*(m)$ and $Z^*(m)$ which are often monotonic. $\Phi^{(n)}$ is expressed in terms of $Y^{(n)}$ and $Z^{(n)}$ and also characterises a property of the nth component of the composite. Despite the monotonic character of $Y^*(m)$ and $Z^*(m)$, the effective property Φ^* that combines characteristics from $Y^*(m)$ and $Z^*(m)$ exhibits a non-monotonic $\Phi^*(m)$ dependence. This is associated with the effect of the combination of the properties [7, 8, 32]. This effect is important, for example, when considering the piezoelectric coefficients

$$g_{3j}^* = d_{3j}^*/\varepsilon_{33}^{*\sigma} \quad \text{and} \quad h_{3j}^* = e_{3j}^*/\varepsilon_{33}^{\xi} \qquad (2.2)$$

($j = 1$ and 3) of the FC/polymer composites [7, 26, 32] with connectivities such as 1–3, 0–3 or 2–2. For the piezoelectric coefficients, g_{33}^* and h_{33}^* of these composites, one can find relatively wide volume-fraction (m) ranges [7, 26, 29, 31, 32] where conditions $g_{33}^* > g_{33}^{(1)}$ and $h_{33}^* > h_{33}^{(1)}$ hold [here, superscript '(1)' refers to the FC component].

The example of the non-monotonic volume-fraction dependence of the piezoelectric coefficient g_{33}^* of the 1–3 FC/polymer composite is shown in Fig. 1.1b. This is achieved at a rapid increase of the piezoelectric coefficient d_{33}^* (Fig. 1.1a) and at a slow increase in the dielectric permittivity $\varepsilon_{33}^{*\sigma}$ in the range of the relatively small volume fractions of FC, i.e. when a contribution from the large dielectric permittivity of FC $\varepsilon_{33}^{(1),\sigma}$ in $\varepsilon_{33}^{*\sigma}$ is small. We remind the reader that g_{33}^*, d_{33}^* and $\varepsilon_{33}^{*\sigma}$ are linked in accordance with (2.2). Thus, in the presence of the polymer component, we can reach the higher piezoelectric sensitivity of the two-component composite in comparison with its poled FC component.

Of independent interest are studies on the FC particle-size dependence of ferroelectricity in the FC/polymer composites. Experimental results [33] show that this dependence can be considerable, and the properties of several FC inclusions are to be taken into consideration to estimate some effective parameters of the piezo-active composites.

2.3 Role of Electric Poling

Poling conditions for the FC/polymer composite differ from the poling conditions for its FC component because of the presence of the polymer component with properties that strongly differ from those of FC, such as dielectric permittivity for example [1]. Many characteristics of the piezo-active matrix composites [20] depend on electric poling conditions including temperature of poling and the features of the electric-field distribution in samples. Due to the lower electric resistivity and large dielectric permittivity of the FC component in the FC/polymer composite, the electric flux concentrates more effectively on the polymer component (that is the matrix in the 0–3 composite) and does not affect the FC inclusions (isolated particles in the polymer medium) to a large extent. At large volume fractions of FC in such a matrix composite, the thickness of a polymer layer that separates the FC inclusions becomes smaller. This facilitates the electric poling of the composite sample as a whole. However, at smaller volume fractions of FC (as a rule, less than 50%), the applied external electric field is imposed on the polymer matrix rather than on the FC inclusions therein. This circumstance is to be taken into account when trying to improve the electromechanical properties of the FC/polymer composites.

A theoretical description of the poling conditions and examples of the effective electromechanical properties of the 0–3 FC/polymer composites at large volume fractions of the PZT-type FC component are considered in work [34]. According to the results [34], the piezoelectric coefficient d_{33}^* of the poled 0–3 composite is proportional to the volume fraction of FC, poling ratio, local field coefficient and the piezoelectric coefficient $d_{33}^{(1)}$ of FC. The poling ratio and local field coefficient are concerned with the shape of the FC inclusion and depolarising field in the composite sample. It should be noted that the predicted effective properties of the studied 0–3 composite and its poling conditions are consistent with experimental data [34]. The concepts developed in work [34] are also in agreement with experimental and theoretical results obtained by other authors, see, for instance, work [35–37].

In comparison with the 0–3 connectivity pattern, the 1–3 connectivity pattern (Table 1.1) is more suitable for electric poling due to the system of aligned continuous FC rods with remanent polarisation vectors $\boldsymbol{P_r}$ oriented parallel to the electric field $\boldsymbol{E}$. This leads to a higher piezoelectric activity and sensitivity of the 1–3 FC/polymer composite in comparison with its 0–3 analogue, see, for instance, data in Fig. 1.1. However, a higher cost of the manufacturing of the 1–3 composites can be a reason for an application of a combined approach [20] developed by Khanbareh. This approach includes combining FC powder with a polymer component and allows a mass production of cost-effective FC/polymer composites. It is obvious that many technological aspects and ways to improve the piezoelectric properties of such composites are linked with poling conditions. The electric field E and temperature T at poling are to be chosen in certain ranges to avoid a considerable heating of the sample and softening its polymer component, as well as to avoid dielectric breakdown phenomena in the sample. The poling electric field E is also to be chosen by taking into account a depolarising field [1] caused by a FC inclusion in a FC/polymer medium that surrounds this inclusion. We remind the reader that the depolarising field can be decreased [26, 34] at the prolate shape of the FC inclusion with the remanent polarisation vector $\boldsymbol{P}_r \uparrow\uparrow \boldsymbol{E}$. The exposition of the field E on poling is to be sufficient for charge transportation over the composite sample and for a formation of a stable charge distribution at a specific arrangement of remanent polarisation vectors $P_{r,i}^{(1)}$ of the FC inclusions therein. We note that the electrical conductivity of the components also strongly influences the electromechanical and pyroelectric properties of the matrix composites based on FCs [6, 8, 38–42]. For example, an enhanced conductivity of the matrix can shorten the build-up time of the electric field acting on the FC inclusions, and this leads to a higher poling degree of the composite sample. As a consequence, considerable improving the piezoelectric and pyroelectric properties of the composites is stated [38, 43]. A reliable way to improve the conductivity of the matrix in the composite sample is to introduce a small volume fraction of a third component with a considerable electric conductivity [44–46].

An intermediate state between the 0–3 connectivity pattern (particulate composite) and 1–3 connectivity pattern (fibrous composite) is often achieved by means of the DEP treatment on a semi-dilute solution of FC particles in a viscous matrix; this is typical of many thermosetting polymer systems prior to curing [9, 10, 19, 20]. In this case, a local electric field on the FC particles leads to an optimum orientations of these particles with respect to the external field $\boldsymbol{E}$ and, as a consequence, leads to the formation of thread-like structures. Some features of DEP and its applications to form piezo-active composites are discussed in Sect. 2.4.

2.4 Dielectrophoresis, Its Characteristics and Advantages

2.4.1 Dielectrophoresis Processing: Application to Composites Based on Ferroelectric Ceramics

The DEP method is of importance at the formation of the FC/polymer composite with a system of prolate inclusions that become oriented along the poling electric field $\boldsymbol{E}$ and promote the improved piezoelectric performance of the composite [9, 19, 20].

During processing of the composite [20], FC particles are dispersed in the resin component of the epoxy and mixed at a speed of 1000 revolutions per minute (RPM) for 15 min using a planetary mixer (SpeedMixer DAC 150.1 FVZ, Hauschild). A hardener is then added and the composite resin is mixed further at 1500 RPM for 3 min. The obtained uncured FC/polymer mixture is degassed and poured into a mould consisting of a 1-mm-thick Teflon sheet with 15 mm diameter circular cutouts. The sheet is then placed between two layers of the aluminium foil (50 μm thick), and this foil serves as a temporary electrode for the application of the electric field $\boldsymbol{E}$, see the schematic of the mould in Fig. 2.1. This electric field causes DEP in the composite sample.

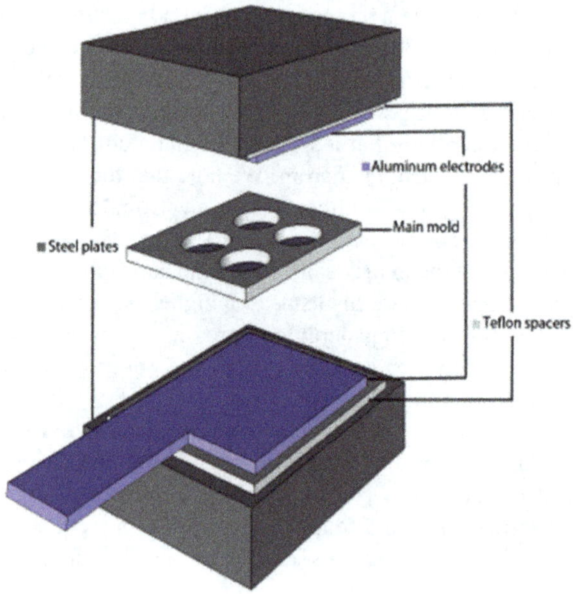

Fig. 2.1 Schematic of the mould used at DEP processing of composites (reprinted from paper by Khanbareh et al. [10], with permission from IOP Publishing)

The set-up as a whole is placed between two bolted steel plates with additional Teflon spacers that separate the electrodes from the steel plates to apply pressure and produce the flat composite samples. Structuring of the FC particles in the yet uncured composite is implemented during the DEP process when an electric field $E = 2$ kV/mm is applied at a frequency $f = 1200$ Hz to the composite sample. Hereby, a function generator (Agilent, 33210A) coupled to a high-voltage amplifier (Radiant Technologies Inc., T6000HVA-2) is utilised at room temperature conditions for 1 h (i.e. well into the curing stage) followed by 5 h heating at $T = 323$ K of the composite to obtain a fully cured composite matrix. The peak-to-peak output voltage of the high-voltage amplifier, the phase angle and the leakage current are verified with an oscilloscope (Agilent, DSO-x 2004A). The randomly dispersed samples are obtained in a similar manner, simply without applying the structuring electric field and by oven curing at $T = 323$ K for 5 h. The completely cured disc-shaped composite samples of 15 mm diameter and 1 mm thickness are polished to remove the upper epoxy layer and postcured at $T = 373$ K for 1 h to remove moisture. After this curing stage, gold electrodes of 9.2 mm diameter and 50 nm thickness are deposited on both sides of the composite sample by sputtering (Balzers Union, SCD 040). The electroded composite samples are poled at $E = 9$ kV/mm and $T = 353$ K in a water-cooled oil bath (Julabo, SE Class III, 12876) within 1 h. The samples are then cooled to room temperature in the poling field $\boldsymbol{E}$.

In order to remove the injected charges during the poling stage and the trapped charges due to impurities, the composite samples are stored at $T = 373$ K for 24 h with their electrodes that become short-circuited prior to further measurements of the physical properties and other parameters of the manufactured composite.

2.4.2 *Effect of Dielectrophoresis Structuring on Electric Displacement and Electromechanical Strain*

The effect of the DEP structuring on the microstructural characteristics of the FC/polymer composite is shown in Fig. 2.2. The clear difference is observed between the microgeometry of the non-structured (or random) and structured composite samples at the volume fraction of FC $m = \text{const}$. Due to DEP, the FC particles (inclusions) construct chain-like structures in the composite sample along the vertical electric field $\boldsymbol{E}$ direction, see Fig. 2.2d. The degree of their alignment that strongly influences the effective physical properties of the composite depends on force acting on the FC particles in the composite. The physical nature of this force is caused by the imposed electric field $\boldsymbol{E}$. This driving force behind the attraction between the particles is a function of the amplitude and frequency of the electric field $\boldsymbol{E}$, the dielectric properties of the FC and polymer components and the size of the FC particles [9, 20]. Although the gravitational force also affects the motion of the FC particles in the

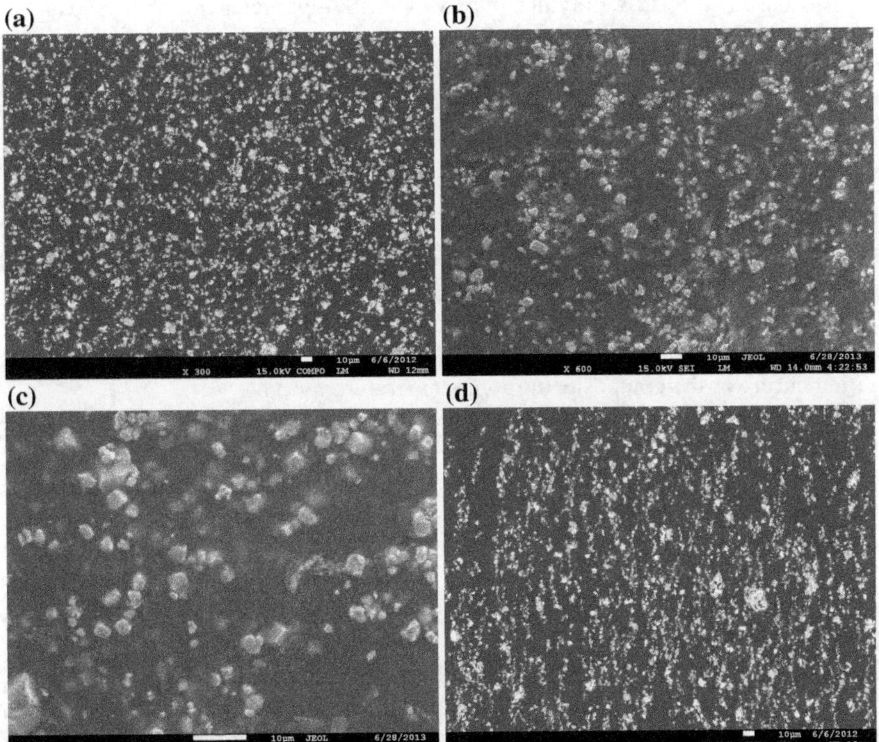

Fig. 2.2 Micrographs of non-structured (**a**–**c**) and DEP-structured (**d**) PT FC/polymer composites at the volume fraction of PT FC $m = 0.2$

fluid polymer medium, this force is not taken into account since no sedimentation is observed in the composite microstructures (see, for instance, Fig. 2.2).

The effect of the DEP structuring is discussed based on an experimental study [20] of a piezo-active composite that is based on a PZT-type FC. The main component of this composite is a PZT5A4 FC powder that is a niobium-doped PZT FC powder with a Zr/Ti molar ratio of 0.52/0.48 (Morgan Electroceramics, UK), i.e. the PZT composition is located close to the morphotropic phase boundary [47–49] of the phase diagram. In the initial stage of the composite manufacture, this powder is calcined at $T = 1423$ K for 1 h according to the optimised scheme put forward by van den Ende [50]. The agglomerated powder is then dry-milled, using 5-mm zirconium balls for 2 h in a single G90 jar mill. The particle-size distribution of milled powder in an aqueous solution with 10% isopropyl alcohol, measured by a Beckman Coulter LS230 laser diffraction analyser, suggests that the average size of the particle is 2.5×10^{-6} m. In order to avoid a moisture adsorption, this powder is stored in a drying oven at $T = 393$ K for 24 h prior to the experimental study. A two-component epoxy system Epotek 302-3 M [Epoxy (diglycidyl ether of bisphenol-A (DGEBA) resin and poly(oxypropyl)-diamine (POPD) multifunctional aliphatic amine curing agent] is used to form a polymer matrix. The PZT particles are dispersed in the resin component of the epoxy system to achieve the specific volume fractions of the FC component ($m = 0.1$, 0.2, 0.3, 0.4 and 0.5) and then mixed at a speed of 1000 RPM for 15 min, using a planetary mixer (SpeedMixer DAC 150.1 FVZ, Hauschild). Subsequently, the hardener is added, and the composite resin is again mixed at 1500 RPM for 3 min. The uncured ceramic/polymer mixture is then degassed and poured into a mould consisting of a 1-mm-thick Teflon sheet with circular 15 mm diameter cut-outs.

The obtained composite is cured at room temperature. To produce a structured 0–3 composite, the slurry is placed between two layers of 50 μm thick aluminium foil. This foil plays the role of the temporary electrodes at DEP, see Sect. 2.4.1. The DEP structuring is performed using an electric field $E = 2$ kV/mm at $f = 2$ kHz during curing of the composite at room temperature for 3 h.

The dielectric properties of the composite are studied using the parallel plate capacitor method with an Agilent 4263B LCR metre (Japan) at a voltage $U = 1$ V and $f = 1$ kHz. The longitudinal piezoelectric coefficient d_{33}^* is measured with a high precision PM300 PiezoMeter (Piezosystems) at $f = 110$ Hz and under a static and dynamic force of 10 N and 0.25 N, respectively. At least five samples for the composite system at $m =$ const are tested. Ferroelectric polarisation measurements are performed using a Radiant technologies Precision Multiferroic Test System at $f =$ 1 Hz and under a double bipolar triangular pulse, with a peak from 0 to 18 kV/mm. Out-of-plane displacement measurements are carried out using a photonic sensor, MTI-2100, with a 2032X high-resolution module in combination with the Precision Multiferroic Test System [20]. The sensor utilises adjacent pairs of light-transmitting and light-receiving fibres. It operates by measuring the interaction between the field of illumination of the transmitting, or source fibres, and the field of view of the receiving, or detector fibres. As the surface of the sample moves away from the sensor tip, the signal decreases, and as the surface moves closer to the sensor tip, the signal increases. The probe is used in the high-resolution mode with a sensitivity of 1.9×10^{-8} m/V. The longitudinal strain ξ_3 is measured on the composite samples at $f = 1$ Hz under double bipolar alternating electric field E using the relation $\xi_3 = \Delta\, t/t$ where t is the thickness of the sample under test, and Δt is the change in the thickness of the sample at the converse piezoelectric effect.

Examples of the electric displacement and strain behaviour of the PZT5A4 FC at $f = 1$ kHz are shown in Fig. 2.3. As follows from measurements at room temperature, the maximum polarisation of the FC samples reaches a value of 0.42 C/m^2, a remanent polarisation of the FC samples is 0.38 C/m^2, and a coercive field $E_c = 1.3$ kV/mm is achieved [20]. The effect of the DEP structuring on the electric displacement and bipolar strain of the composite along the poling direction is shown in Figs. 2.4 and 2.5, respectively. The results shown in Figs. 2.4 and 2.5 are obtained at a comparison of a stimulus (electric field $\boldsymbol{E}$ vector being parallel to the poling axis OX_3) and response (either electric displacement D_3 or mechanical strain ξ_3) on the poling direction OX_3. It is seen that in the DEP-structured composite the larger D_3 (Fig. 2.4) and ξ_3 (Fig. 2.5) values are achieved at $E_3 =$ const. As follows from experimental results [20], the D_3 and ξ_3 values at $E_3 =$ const increase in increasing the volume fraction of FC m in the structured composite. This means that the system of the FC particles arranged mainly along the $\boldsymbol{E}$ vector promotes the better dielectric and piezoelectric characteristics of the structured composite in comparison with its non-structured analogue with a random distribution of the FC particles in the polymer medium.

The effect of the DEP alignment of the FC particles on the electromechanical properties of the piezo-active composite is analysed [10] in terms of the MCM [51]. This model enables one to estimate a fraction of the 1–3 connectivity in the studied composites and, therefore, an influence of this connectivity on the piezoelectric properties and related parameters of the composite. Within the framework of this model, both parallel- and series-connected regions are distributed over the piezo-particulate composite sample. The 1–3 connectivity in the composite is caused by a direct contact of particles spanning the thickness of the sample in the poling direction [20, 51]. Results of the application of the MCM to the PZT-based composite are shown in Fig. 2.6a. Data in Fig. 2.6 illustrate the key role of the 1–3 connectivity in forming the high level of the piezoelectric longitudinal strain in the structured composite sample.

Further examples of the improved performance of the FC-based composites manufactured by means of DEP will be discussed in Chap. 3.

2.4.3 Reproducible Composite Patterns and Related Microgeometric Features

When manufacturing the piezo-particulate FC/polymer composites by means of DEP, a problem of reproducibility of the composite patterns appears. It is obvious that, due to links between the microgeometry and properties of the piezo-active composite,

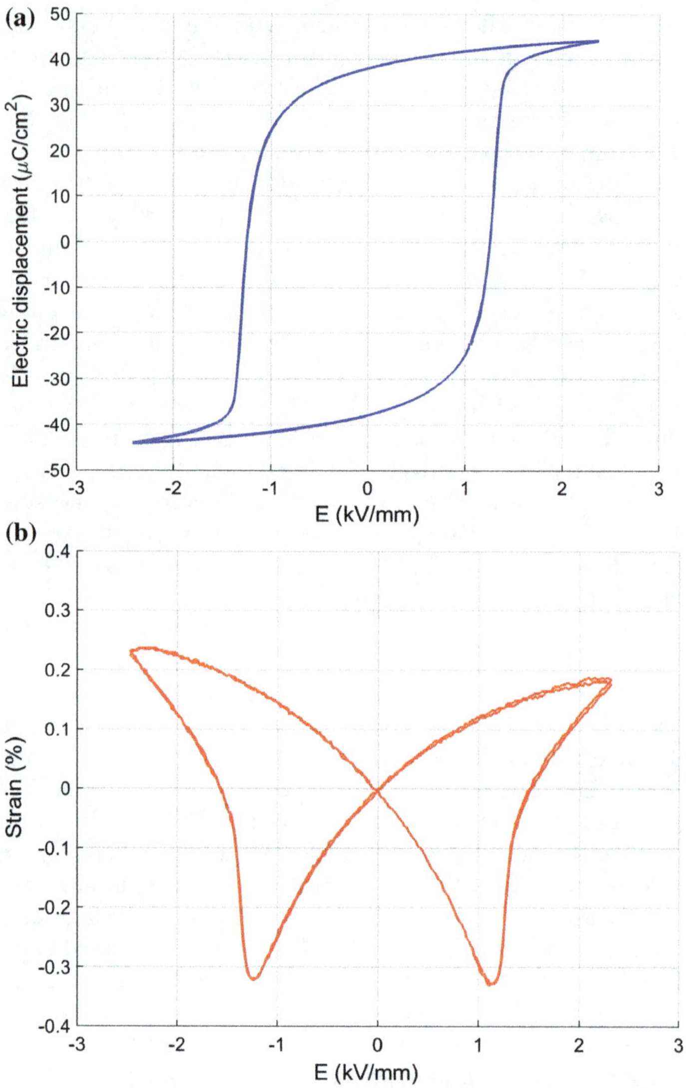

Fig. 2.3 Hysteresis loops of the PZT5A4 FC at 1 kHz and room temperature: **a** electric displacement—electric field [$D_3(E_3)$] loop and **b**, mechanical strain—electric field [$\xi_3(E_3)$] loop [20]

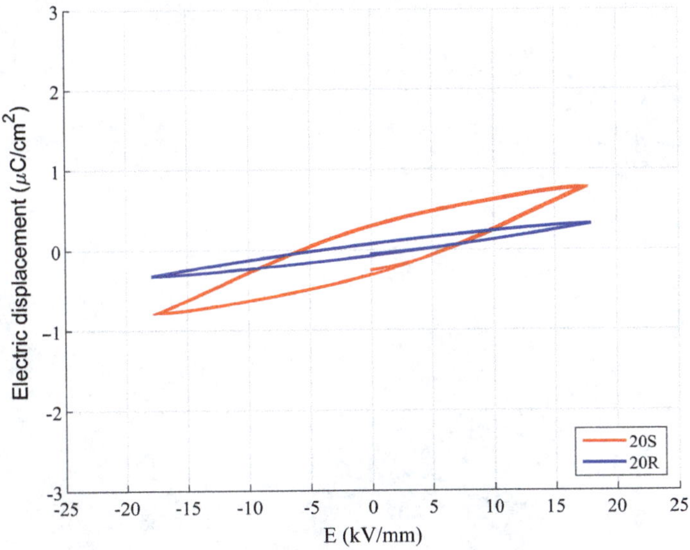

Fig. 2.4 Effect of DEP structuring on the electric displacement—electric field loop for the PZT5A4 FC/epoxy composite at $m = 0.2$. R and S are related to the random and structured composites, respectively [20]

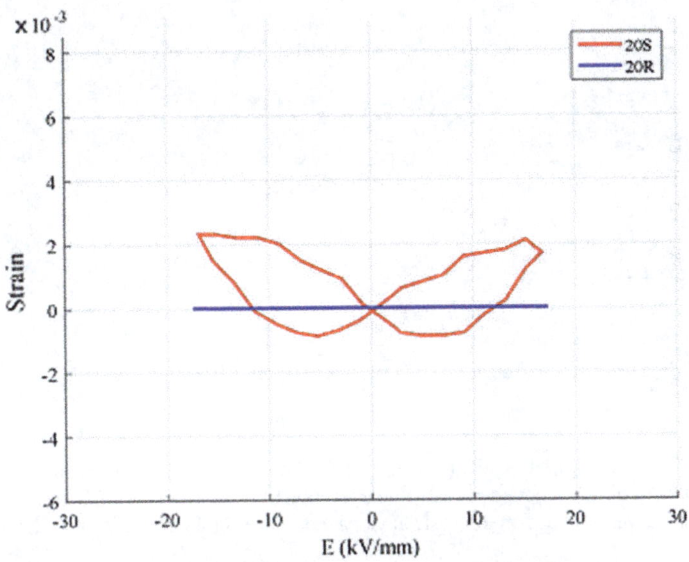

Fig. 2.5 Effect of DEP structuring on the mechanical strain—electric field loop for the PZT5A4 FC/epoxy composite at $m = 0.2$. R and S are related to the random and structured composites, respectively [20]

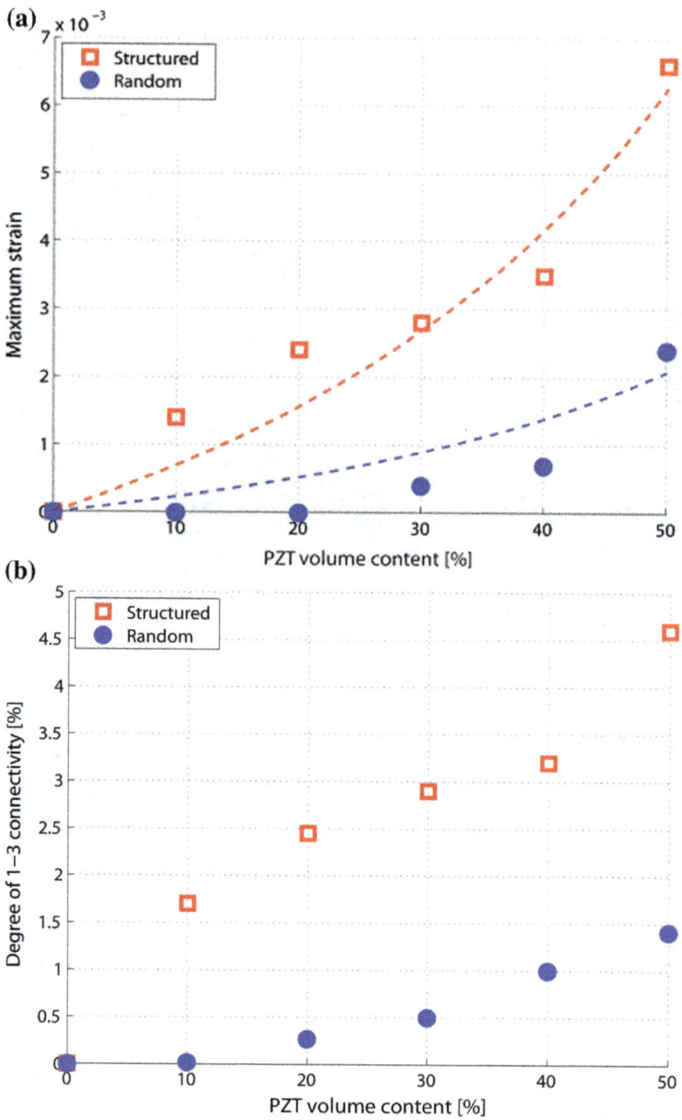

Fig. 2.6 Maximum longitudinal strain at $E_3 = 18$ kV/mm and $m = \text{const}$ (**a**) and fraction of the 1–3 connectivity (**b**) as a function of the volume fraction of FC in the PZT FC/epoxy composite. In graph a, dash curves are built on the basis of calculations within the framework of the MCM [20]

the manufacturing procedure is to be developed for obtaining the composite samples with stable and well-reproducible properties. The DEP method enables us to obtain the relatively stable microgeometry in the composites wherein the 0–3 and 1–3 connectivity patterns coexist (see, for instance, Fig. 2.2d), and this argument is important at the further description of the properties of these materials.

It is important to attain the appreciable fraction of the 1–3 connectivity and to prevent destroying the fibrous composite system in the sample. We remind the reader that the 1–3 connectivity patterns mean the continuous distribution of the FC component (rod, fibre or chain of particles) along the poling direction [7], and such a distribution improves the longitudinal piezoelectric effect (piezoelectric coefficient d_{33}^*) and influences the piezoelectric anisotropy concerned with the d_{33}^*/d_{31}^* ratio. The 0–3 connectivity pattern influences both the d_{33}^* and d_{33}^*/d_{31}^* values, however, to a lesser extent [7, 32] in comparison with the 1–3 connectivity pattern due to the isolated character of piezoelectric (or FC) particles in the 0–3 composite. Specifics of poling conditions by taking into account electrophysical parameters and microstructure of the FC component (including the FC grain shape, texture and alignment with respect to the poling axis), selection of the FC and polymer components and temperature range for the exposition of the electric field are main factors that can strongly influence the microgeometry and effective physical properties of the DEP-structured composite.

2.5 In Situ Structuring and Poling

The key parameters controlling the effective electromechanical properties of the structured composites are the quality of particle alignment and the poling efficiency [20, 52]. Improving the filler orientation (i.e. the orientation of the FC chains in the composite sample) at specific low volume fractions of FC while leads to a high piezoelectric sensitivity [7] and considerable flexibility of the composite sample [19, 53–55]. Enhancing the alignment quality at decreasing interparticle distance in the composite sample leads to improvements in the dielectric, piezoelectric and pyroelectric properties of the piezo-particulate composite [10]. In conventional FC/polymer composites, the polymer component is characterised by a smaller dielectric permittivity in comparison with the FC component (see e.g. Tables 1.2 and 2.2) and by a lower conductivity [56] as compared to the conductivity of FC. Thus, a strong AC field at elevated temperatures for an extended period is required for efficient poling the composite sample [35, 56]. This means that the poling process and effective properties being formed in the composite remain a challenge in the manufacturing field. Results of recent studies [10] suggest that the conductivity of the both FC and polymer components strongly influences the effective properties of the piezo-particulate composite. It should be noted that an enhanced electrical conductivity of the polymer matrix in such a composite shortens the build-up time of the electric field acting on the FC particles therein. As a consequence, in the case of a matrix with a large conductivity, an efficient poling of the composite sample can be achieved even at room temperature, at relatively short poling times and relatively weak electric fields. The experimental results show that the conductivity of thermoset polymer matrices as well as their dielectric permittivity decrease upon curing. Thus, combining the DEP structuring in the AC electric field and DC poling conditions is possible, while the matrix remains in the liquid state, which can improve the poling efficiency of the composite. Moreover, the structuring and poling technique leads to a cost-effective single-step manufacturing of the FC/polymer composites. In work [52], the in situ DEP structuring and poling process of the piezo-particulate PZT FC/epoxy composites in the AC and DC electric fields is highlighted. The effect of processing parameters on the structuring configuration and on the final effective properties of the composites is described.

To manufacture the DEP-structured composite, PZT5A4 powder (Nb-doped PZT powder with a Zr/Ti molar ratio of 52/48) from Morgan Electroceramics, UK, is first calcined at $T = 1423$ K for 1 h according to the optimised scheme [50]. The agglomerated powder mass is then dry-milled using 5-mm zirconium balls for 2 h, using a single G90 jar mill. The particle-size distribution of milled powder in an aqueous solution with 10% isopropyl alcohol is measured by a Beckman Coulter LS230 laser diffraction analyser. According to experimental results on the size distribution, the particle size is characterised by $d(10) = 1.2$ μm, $d(50) = 2$ μm and $d(90) = 8.5$ μm. The powder was stored in a drying oven at $T = 393$ K for 24 h prior to the experiment, to avoid a moisture adsorption. As for polymer, a two-component epoxy system [Epotek 302-3M, Epoxy (diglycidyl ether of bisphenol-A resin) and poly(oxypropyl)-diamine multifunctional aliphatic amine curing agent] is used. Both of these polymers are degassed separately in a vacuum oven at room temperature for 3 h prior to the experiment to avoid an air void formation in the polymer medium. The PZT5A4 FC particles are dispersed in the polymer resin to the specific volume fractions ($m = 0.1$, 0.2, 0.3, 0.4 and 0.5). Subsequently, the slurry is degassed for 5 min, then mixed with the hardener and degassed within next 5 min. After this procedure, the slurry is moulded. The details of the mould layup are given in work [10].

Three sets of composite samples are then prepared at specific combinations of AC and DC bias fields, and the parameters of the external electric fields $\boldsymbol{E}$ applied to the samples are shown in Table 2.3. The electric field is applied using a function generator (Agilent, 33210A) coupled to a high-voltage amplifier (Radiant Technologies Inc., T6000HVA-2). The optimum frequency level is obtained for the highest phase angle of Lissajous plots for each volume fraction of FC m = const in early stages of the cure. Each set of the samples is exposed to the specific electric field for 3 h, then cured overnight at room temperature in the presence of the electric field. The samples are stored at $T = 373$ K for 24 h. The electrodes on the samples are short-circuited prior to the measurements in order to remove injected charges during the poling stage and trapped charges due to impurities.

As follows from experimental data, the first two sets mentioned in Table 2.3 result in quasi 1–3 composites with different degrees of structuring, and set 3 leads to a 0–3 composite poled at curing. The properties of the manufactured FC/polymer composites are compared to the properties of the reference samples from work [19].

Table 2.3 Different combinations of electric fields ***E*** used for manufacturing of the FC/polymer composites

Set	E_{AC} (kV/mm), AC field	f_{AC} (kHz)	E_{DC} (kV/mm), DC field
1	1	2	2
2	0.1	2	2
3	0	0	2

In the second column, the amplitude values of the AC field are shown

It should be noted that the reference composite samples are processed at an applied electric field with an amplitude E_{AC}= 1 kV/mm and f = 4 kHz, poled at E = 10 kV/mm and T = 373 K in a silicone oil bath within 30 min [19]. Microstructures of the composite samples are observed by means of a field-emission scanning electron microscope (JEOL, JSM-7500F) that is operated in the backscattered electron mode. The samples sectioned parallel to the formed particle chains are embedded in curing epoxy at room temperature and polished with diamond paste. The average size of diamond particles in this paste is approximately 1 μm.

Figure 2.7 shows the scanning electron microscope (SEM) images of the PZT FC/epoxy composites at m = 0.2. These composites have been manufactured by means of conventional method and in situ structuring and poling (see parameters in Table 2.3). The average particle size defined from the SEM microstructures is in agreement with our result from the particle-size analysis. In Fig. 2.7, we observe some chain-like structures that are formed along the ***E*** direction during DEP. In the random composite samples, the PZT FC particles are homogeneously distributed in the epoxy matrix, see set 3 and a reference random postcuring-poled composite sample. In a case of set 1 (at the application of a strong AC field), we state a well-developed fibrous structure of the composite, and this resembles the structure of a reference quasi 1–3 composite sample that has been produced by successive DEP and poling process steps. Application of a lower AC field strength (set 2) leads to a system of well-oriented FC particles. However, in this case a less developed fibrous structure, with wider particle-free regions separating the chains oriented along the ***E*** direction is observed. As follows from the analysis of the degree of orientation of the systems of FE particles in the studied composite samples [20], a slightly higher degree of orientation is observed for set 1, compared to set 2. This is concerned with a larger driving force, acting on the FE particles, in response to the larger amplitude of the AC electric field E applied for structuring [57].

2.6 Manufacturing of Piezo-Active Composites for High-Temperature Applications

Piezoelectric materials suitable for applications in automotive, aerospace, energy harvesting and related branches have been of great interest in recent decades [58].

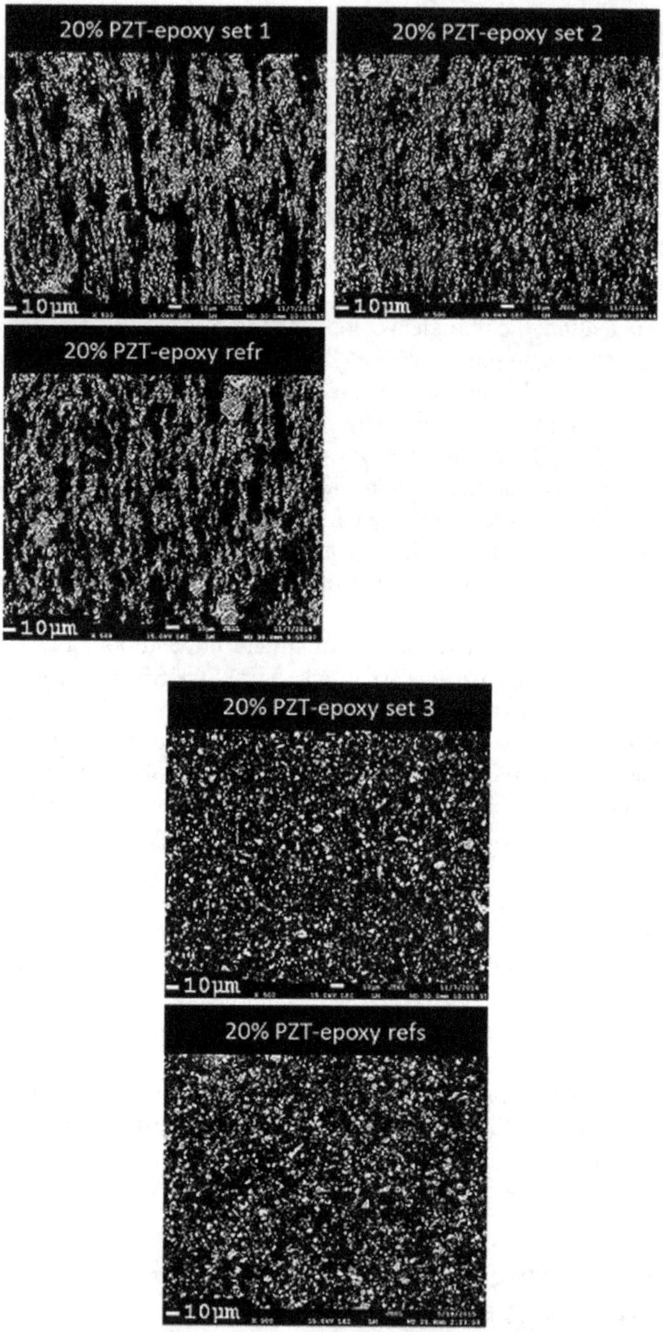

Fig. 2.7 SEM micrographs of PZT/epoxy composites prepared using conventional and in-situ techniques [20], three sets at the volume fraction of PZT $m = 0.2$. In micrographs 'refr' and 'refs' are given to denote reference random (or non-structured) and reference structured samples, respectively

These materials are often exploited in a high-temperature range, at $T > 400$ K. Examples include modern automotive electronics, where repeated thermal cycles take place at $T \approx 400$–420 K, and composite-based devices could operate during thermal cycles with upper temperatures of $T \approx 700$–800 K in the future. In the aerospace sector, structural health monitoring of the engines requires piezoelectric sensors that are to be located as close as possible to an engine, and this undoubtedly leads to the exploitation of sensors at high temperatures.

The exploration of important clean-energy sources, such as nuclear and wind power, as well as geothermal and electrical energy, has stimulated a development of high-temperature electronic materials in the last decades. High-temperature sensors for structural health monitoring can ensure an enhanced lifetime and reliability, less maintenance and downtime and, therefore, can reduce an overall cost of renewable energy sources [59, 60] Piezoelectrics can be used for energy harvesting that is concerned with a conversion of energy from mechanical into electrical energy [1, 7, 26]. For example, energy harvesting under extreme conditions in oil producing wells, wherein ambient pressures can reach 200 MPa and temperature $T \approx 470$ K, needs a piezoelectric element that remains stable in wide pressure and temperature ranges. In this case, local power production is crucial since transmitting power from the surface of the piezoelectric element is complicated, due to the difficulty of making reliable electrical connections in harsh environments [61].

Conventional FC/polymer composites are applied in high-temperature ranges to a restricted degree because of a reduced sensing capability of the FC component at elevated temperatures, increased electrical conductivity and mechanical attenuation. In addition, a temperature dependence of the piezoelectric properties of the FC component is also to be taken into account. The low glass transition temperature T_g and high thermal expansion of the polymer matrix limit the operational temperature range of the FC/polymer composites.

Of specific interest is a piezo-active composite that contains a FC component with high Curie temperature T_C and a passive or FC high-temperature polymer. For such a composite, its operating temperature range is to be substantially below T_C in order to minimise a thermal ageing and degradation of the physical properties. Among numerous polymer components, amorphous polyimides are important candidates for the aforementioned applications [62–65]. Of particular interest is that viable, multifunctional and lightweight materials for demanding space applications stimulated the development of new series of such polymers [62, 63, 66] with superior mechanical and chemical properties. Some of the polyimide systems exhibit insignificant piezoelectric properties [62, 64] in comparison with the conventional FCs of the perovskite type (see data in Table 1.2). One of the most widely reported piezo-polyimide polymers is a polyimide that contains a single nitrile group, (b-CN)APB/ODPA. The schematic of this organic compound is shown in Fig. 2.8. In contrast to the well-known polyvinylidene fluoride (PVDF) polymer family [3], the high-temperature performance of a thin film made of the aforementioned polyimide enables one to retain 50% of the room temperature remanent polarisation at $T \approx 420$ K [64]. However the remanent polarisation of this polyimide is relatively small ($P_r = 0.014$ C/m^2) after the poling procedure at electric field $E = 100$ kV/mm

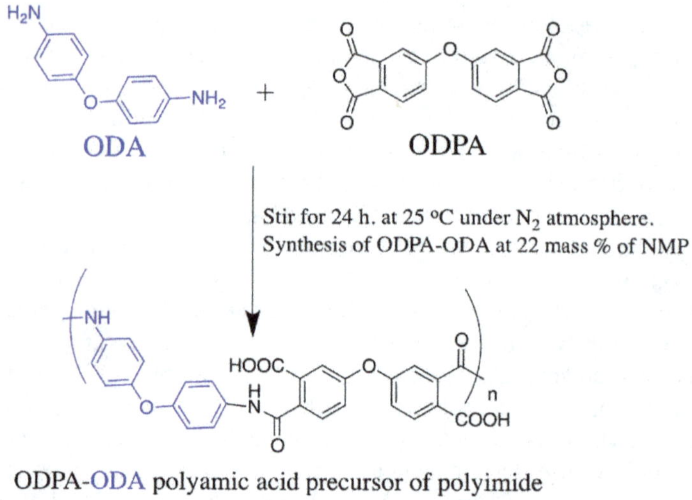

Fig. 2.8 Schematic of the (b-CN)APB/ODPA polymerisation [20]

and temperature $T > T_g$ for 1 h. As a result of such a poling, piezoelectric charge coefficient of this polyimide is much lower than that of PVDF [62, 64]. A limited chain mobility in the film restricts the preferential alignment of the dipoles under the electric field $\boldsymbol{E}$. To solve this problem, some polymers have been synthesised with different types and concentrations of the polar groups [63]. Despite the interesting research outcomes, the polyimide systems exhibit low piezoelectric responses that may restrict their wide application [62, 63]. To overcome these shortcomings and to achieve appropriate combinations of mechanical, thermal and dielectric properties, researchers have manufactured piezo-active composites wherein the FC particles were embedded in a polyimide medium. Structuring the FC particles into a chain form configuration, by means of DEP, leads to an improvement of the dielectric, piezoelectric and pyroelectric properties of the composites compared to the similar properties of random composites (see, for instance [10, 19, 20, 53]).

A processing procedure for the high-temperature piezo-active composites has been put forward by Khanbareh [20]. Among the various potential commercially available FC particles, PT is a good high-temperature candidate with $T_C = 765$ K, a large spontaneous polarisation and relatively small dielectric permittivity [67–70]. The ODA/ODPA polyimide is prepared using a standard 2-step synthesis method via a polyamic acid solution in N-Methyl-2-pyrrolidone, as shown in Fig. 2.8. The thermal properties of the manufactured polyimide film are studied by means of the thermogravimetric analysis using a Perkin Elmer Pyris Diamond, as well as differential scanning calorimetry using a Perkin Elmer Sapphire DSC. Samples are heated at a rate of 20 K/min under a nitrogen atmosphere to $T = 823$ K. Dynamic mechanical thermal analysis is carried out in the tensile mode, at a frequency $f = 1$ Hz and a heating rate of 2.0 K/min.

Ferroelectric PT powder purchased from Sigma-Aldrich is calcined at $T = 1073$ K for 2 h to develop a single-phase $PbTiO_3$ composition. Agglomerated PT powder is then dry-milled using 5-mm zirconium balls for 2 h in a single G90 jar mill. The particle-size distribution of milled PT powder in an aqueous solution with 10% isopropyl alcohol, measured by a Beckman Coulter LS230 laser diffraction analyser is $d(50) = 3$ μm. The PT powder is then stored in a vacuum drying oven at $T = 393$ K for 24 h prior to experiments to avoid a moisture absorption. The dried PT powder is predispersed in the polyamic acid solution in N-Methyl-2-pyrrolidone, stirred for 1 h and added to the polyamic acid and the mixture is then cast on an aluminium foil using a doctor blade, with an initial film thickness of 600 μm. Structuring of the PT particles in the polyamic acid polymer is performed during the DEP process. An electric field $E = 2$ kV/mm at a frequency $f = 2$ kHz is applied to the PT/polymer composite medium by using a function generator (Agilent, 33210A) coupled to a high-voltage amplifier (Radiant Technologies Inc., T6000HVA-2) at room temperature for 1 h in an argon purged oven. Once the structuring of the PT particles is realised, the film is dried at $T = 333$ K for 1.5 h under the same atmosphere and electric field. The randomly dispersed (or non-structured) composite samples are obtained by the method described above, however

(i) without applying the electric field,
(ii) by oven curing at $T = 333$ K for 1.5 h, and
(iii) using argon gas as a surrounding medium.

The free-standing PT-based composite films are then removed and cut to discs of 7 mm diameter. Electrodes are sputtered on the samples, and the composites are poled under the electric field $E = 20$ kV/mm at $T = 423$ K for 30 min. The poled films are then imidised in a vacuum oven at $T = 473$ K and 573 K, each for 1 h. The resulting composite films have a fully imidised chemical structure shown in Fig. 2.9.

A clear difference in the structure between the random and DEP-structured composite samples is shown in Fig. 2.10. Comparing the micrographs in Fig. 2.10, we state that during DEP, the PT particles form chain-like structures along the electric field $\mathbf{E}$ vector, i.e. on the vertical direction from bottom in Fig. 2.10. The degree of alignment of these particles depends on the force acting on the particles, as a consequence of the imposed electric field. This degree of alignment strongly influences the effective electromechanical properties of the composite by analogy with the influence of the poling degree of the $PbTiO_3$-type FCs on their electromechanical properties [71, 72].

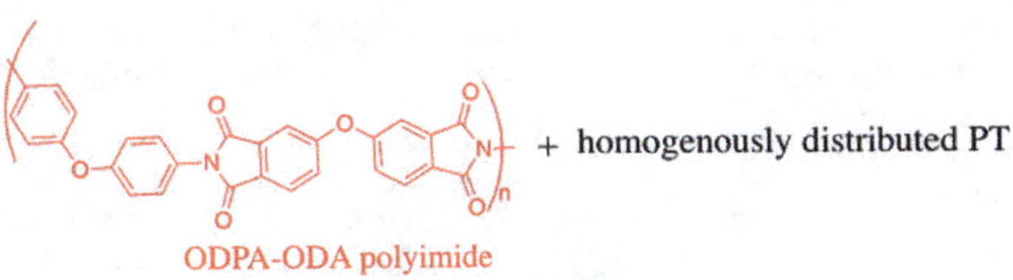

Fig. 2.9 Chemical structure of fully imidised 20% PT-ODPA/ODA composite [20]

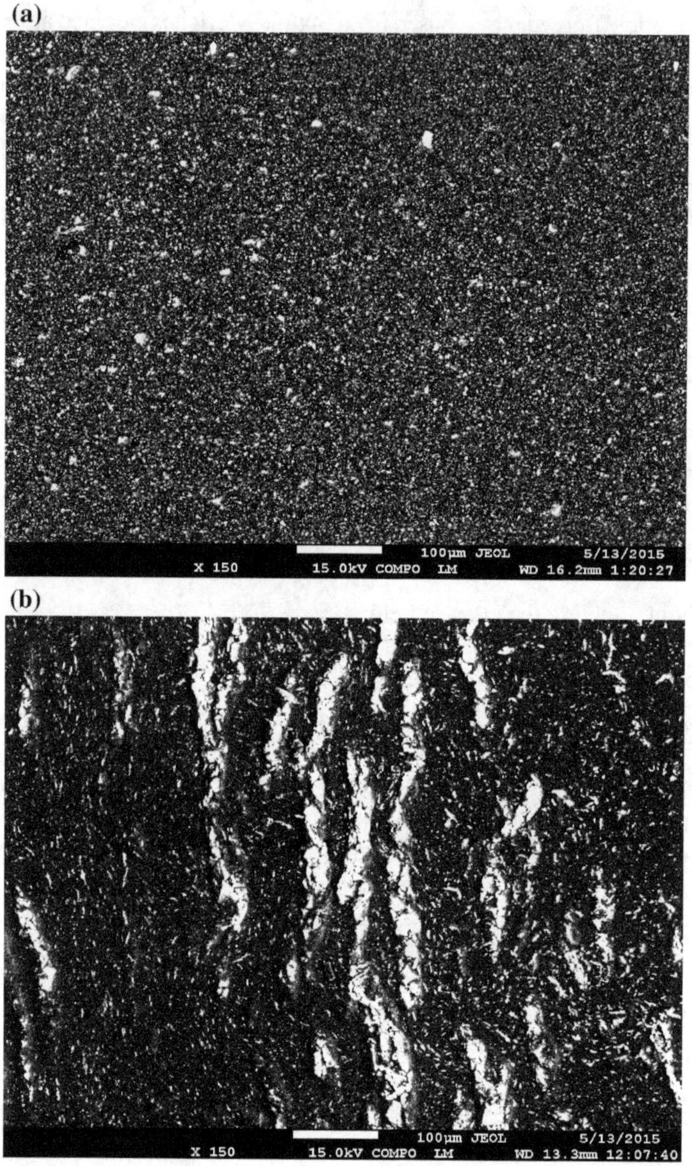

Fig. 2.10 SEM microstructures of the fully imidised PT/ODPA/ODA composite at the volume fraction of PT $m = 0.2$: a, 0–3 random (non-structured) and b, quasi 1–3 structured by means of DEP [20]

2.7 Manufacturing of Piezo-Particulate/Polymer Foam Composites

Polymers can be foamed to achieve a reduction of their density, dielectric permittivity and increase energy dissipation, mass transport and so on [73]. Recently ultra-low permittivity dielectrics have been manufactured for applications in the new-generation microchips that need dielectric inter-layers with the relative dielectric permittivity $\varepsilon < 2.2$ [74]. We remind the reader that, at room temperature, for instance, $\varepsilon = 2.3$ for high-density (monolithic) polyethylene, while $\varepsilon = 3.5$ for polyurethane; see data in Table 2.2. In the last decades, a new concept has been developed that is based on porous dielectric structures. These structures are formed mainly by a thermal decomposition of a block copolymer composed of a thermally stable block and a thermally instable one [75]. Novel porous polymers (or foam polymer) demonstrate a clear relation between its increase in porosity and its decrease in the dielectric permittivity ε [74]. The performance of the polymer foam is determined by taking into account the intrinsic properties of the polymer and its microstructure. As follows from results [76], the dielectric permittivity of a polymer foam decreases linearly in increasing porosity (or a gaseous content). However, in a system that represents homogeneously distributed air inclusions in the polymer matrix, the dielectric permittivity decreases exponentially in increasing the gas content [73, 77]. A polymer foam that consists of randomly distributed closed cells has an effectively reduced dielectric permittivity, and its value is comparable to that related to the dense polymer component.

The DEP structuring of the FC/polymer composite leads to the improved piezoelectric properties, in particular piezoelectric sensitivity, in comparison with the properties of the conventional 0–3 composite with isolated spherical FC inclusions in a polymer matrix. Foaming of the polymer matrix leads to a decrease in the dielectric permittivity of the composite and becomes an important factor that influences the electromechanical properties of the composite [74].

A piezo-particulate/polymer foam composite based on the PZT FC has been manufactured by Khanbareh et al. [74]. PZT5A4 powder [a composition based on $Pb(Zr, Ti)O_3$] from Morgan Electroceramics was calcined at $T = 1423$ K in a closed zirconia crucible according to the optimised scheme by van den Ende [50], and the time of the calcination procedure was 1 h. After this calcination, the dry powder mass was ball-milled in a Gladstone Engineering G90 jar mill for 2 h.

Subsequently, the particles were sieved for 20 min using a Haver & Boecker EML Digital Plus test sieve shaker with stacked sieves with mesh sizes of 125 mm and 63 mm. The powder was then dried for at least 2 h to prevent agglomerations of the FC particles due to moisture, as well as to prevent a chemical interaction between the polymer and moisture from these particles. The particle-size distribution of the milled powder in an aqueous solution with 10% isopropyl alcohol, measured by a Beckman Coulter LS230 laser diffraction analyser, is $d(10) = 0.8$ μm, $d(50) = 1.8$ μm and $d(90) = 6$ μm. A two-component urethane rubber, Smooth-on-Econ 80 Polyurethane, was used to serve as a polymer component in the composite. This polymer is composed of di-isocyanate resin and polyol [74]. Gas formation is observed upon addition of water to the uncured polymer mass, due to a chemical reaction between the di-isocyanates component and water. This reaction leads to the formation of poly(ureaurethane) and the release of CO_2.

To study the effect of demineralised water (D-water) on gas formation in the polyurethane medium, di-isocyanate resin was magnetically stirred at 600 RPM for 6–7 min at room temperature. In the meantime, polyol and a varying amount of D-water (in our experiments from 0 to 0.5%) were mixed using a SpeedMixer DAC 150.1 FVZ at 3500 RPM for 5 min. Subsequently, di-isocyanate resin, polyol and D-water were magnetically stirred at 600 RPM, until the viscosity increased until it was sufficiently high to stop the magnet from rotating. A tape-cast film was made on an aluminium substrate, using a doctor blade that set at 1 mm. The amount of D-water that was added to 10 mL polyurethane is in the mL range.

Thermogravimetric analysis of the manufactured polymer sample was performed using Perkin Elmer Pyris Diamond at the rate of 20 K/min under a nitrogen atmosphere. The results from work [74] show that, at $T = 473$ K, less than 2 mass% loss is registered for this polymer. After thermogravimetric analysis, the differential scanning calorimetry was performed at a heating rate of 20 K/min under a nitrogen atmosphere, using the Perkin Elmer Sapphire to determine the glass transition temperature T_g of the studied polymer. According to experimental data [74], $T_g =$ 281 K, and this temperature does not undergo considerable changes after postcuring at $T = 373$ K. The production route that has been used for the production of the D-water range of 0–0.5% is also suitable [74] for the production of the first batch of the random three-component 0–3-type composites. However, for the case of a volume fraction of PZT FC of $m = 0.20$, the initial viscosity of the heterogeneous system due to the presence of the FC particles is too high to allow for magnetic stirring. This circumstance leads to an appropriate adaptation of the production route [74], where the magnetic stirring step is eliminated.

Polymer foams are able to form nonetheless which proves that the PZT particles can play a role as void nucleation sites, thereby replacing the functionality of microvoids that are introduced by magnetic stirring. To manufacture the composite, the PZT particles are dispersed in the mixture of polymer resin and D-water to the specific volume fractions $m = 0$, 0.20, 0.30 and 0.40 and mixed at 3500 RPM for 1 min. The 0–3-type composite samples are manufactured by means of casting the slurry on an aluminium substrate using a doctor blade at the thickness of 1 mm. The surface of the film was exposed to air. The quasi 1–3 composite samples were manufactured in a closed mould format using a Teflon mould with a thickness 1 mm, see some details of the mould layout in work [10].

For the random and DEP-structured composite samples, the optimum volume fraction of D-water, resulting in the maximum reduction of dielectric permittivity ε of the polymer matrix is found to be 0.4 vol. % and 0.2 vol. %, respectively [74]. It should be noted that the closed mould restricts water evaporation from the mixture, and therefore, a lower water content of 0.2 vol. % results in similar gas content of 60 vol. % as is known for the similar random composites prepared by casting. The DEP structuring is performed on the uncured composite samples by applying an electric field $E = 3$ kV/mm at $f = 3$ kHz for 1.5 h, using a function generator

(Agilent, 33210A) coupled to a high-voltage amplifier (Radiant Technologies Inc., T6000HVA-2), until the polymer matrix of the samples is fully cured. After the curing procedure, flexible films of 1 mm thickness are obtained. The composite samples are poled by using Heinzinger 30,000-5 30 kV DC amplifier, a Haake N3 digital circulating hot oil bath filled with silicone oil and a custom made sample holder at $T = 373$ K. This leads to a poling electric field of $E = 5$ kV/mm, and the poling procedure is 2 h long.

Circular discs of the poled composite are manufactured, and stationary electrodes are made by sputtering gold for 20 min on both sides of each disc using a Quorum Q300T D sputter coater. The electroded samples are punched using a 16 mm punch, thereby removing the material at the edges to prevent an unexpected leakage current. The dielectric permittivity of the composite is measured on Agilent 4263B & 16034E–Inductance Capacitance Resistance Metre (LCR) by the parallel plate capacitor method at a voltage of 1 V and $f = 1$ kHz. The piezoelectric coefficient d_{33} of the poled composite samples was measured using a Piezotest PiezoMeter System PM300—Berlincourt d33 m, under a 10 N static force and a 0.25 N dynamic force at $f = 110$ Hz. To carry out an analysis of composite microstructures, the samples were sectioned using scissors along the thickness, and the cross sections were examined using a field-emission SEM (JEOL, JSM-7500F) [74].

Examples of the performance of the manufactured three-component composites with foam polymer matrices are shown and discussed in Sect. 3.4.

2.8 Conclusion

This chapter has been devoted to the description of methods for manufacturing of modern composites based on ferroelectrics. The overwhelming majority of the methods in this field are concerned with the FC/polymer composites that have been widespread for the past decades, and the main connectivity patterns are 0–3 and 1–3. The methods to manufacture the piezo-active composites are based on knowledge of physics, chemistry, materials science and technological principles. In the present chapter, the main emphasis is placed on applications of DEP to form long chains of aligned and closely connected FC particles in an electric field, when the matrix polymer is in the low-viscosity state.

When manufacturing the particulate FC/polymer composites, two processing routes can be tested and compared. The first route involves using the conventional DEP structuring to form chain-like particle configurations, and this stage of manufacturing is followed by curing the matrix and poling in a high-temperature range and under a high voltage. For the second route, a simultaneous DEP and poling procedures is implemented at room temperature while the polymer is in the liquid state followed by subsequent curing. Hereby, a significant improvement in piezoelectric properties of quasi 1–3 composites is achieved due to the combination of the DEP alignment of the FC particles and the poling process. The obtained composites contain aligned long FC chains which are characterised by elements of 0–3 and 1–3 connectivity patterns, and this feature is to be taken into account in subsequent studies of the characterisation and prediction of the effective properties. Comparison of microstructures and orientations of the FC particles is drawn for three sets of the composite samples that have been prepared at specific combinations of AC and DC bias fields (see data in Table 2.2), and the role of the electric-field structuring (DEP) in forming the specific composite microstructures (Fig. 2.7) is shown.

The processing procedure suitable for the high-temperature piezo-active composites enables one to manufacture the quasi 1–3 FC/polymer samples with high T_C and T_g of the ferroelectric and polymer components, respectively. Hereby, DEP plays the key role in forming the long chains that consist of the large number of FC particles poled under the applied electric field $\boldsymbol{E}$. The alignment degree of the FC particles in the polymer medium depends on the force acting on these particles, and this force is a result of the electric field applied to the composite sample.

Finally, progress in the manufacturing of the piezo-active DEP-structured composite with the foam polymer matrix is observed. This is an example of the three-component composite wherein air plays an important role in forming the electromechanical properties of the composite. The combination of the DEP structuring and porosity in the polymer matrix of the three-component composite and subsequent poling lead to stable properties of the composite as a whole. The methods and technological 'know-hows' to form the similar advanced composite structures can be taken into account when manufactured novel materials for piezo-technical and related applications.

References

1. T.R. Gururaja, A. Safari, R.E. Newnham, L.E. Cross, Piezoelectric ceramic-polymer composites for transducer applications, in *Electronic Ceramics: Properties, Devices, and Applications*, ed. by M. Levinson (Marcel Dekker, New York and Basel, 1988), pp. 92–128
2. V.F. Janas, A. Safari, Overview of fine-scale piezoelectric ceramic/polymer composite processing. J. Am. Ceram. Soc. **78**, 2945–2955 (1995)
3. E.K. Akdogan, M. Allahverdi, A. Safari, Piezoelectric composites for sensor and actuator applications. IEEE Trans. Ultrason. Ferroelectr. Freq. Control **52**, 746–775 (2005)
4. A. Safari, E.K. Akdogan, Rapid prototyping of novel piezoelectric composites. Ferroelectrics **331**, 153–179 (2006)
5. F. Wang, C. He, Y. Tang, X. Zhao, H. Luo, Single-crystal $0.7Pb(Mg_{1/3}Nb_{2/3})O_3$–$0.3PbTiO_3$/epoxy 1–3 piezoelectric composites prepared by the lamination technique. Mater. Chem. Phys. **105**, 273–277 (2007)
6. X. Dongyu, C. Xin, H. Shifeng, Investigation of inorganic fillers on properties of 2–2 connectivity cement/polymer based piezoelectric composites. Constr. Build. Mater. **94**, 678–683 (2015)
7. V.Yu. Topolov, C.R. Bowen, *Electromechanical Properties in Composites Based on Ferroelectrics* (Springer, London, 2009)
8. R.E. Newnham, Molecular mechanisms in smart materials. Mater. Res. Soc. Bull. **22**(5), 20–34 (1997)
9. S.A. Wilson, G.M. Maistros, R.W. Whatmore, Structure modification of 0–3 piezoelectric ceramic/polymer composites through dielectrophoresis. J. Phys. D Appl. Phys. **38**, 175–182 (2005)

10. H. Khanbareh, S. van der Zwaag, W. Groen, Effect of dielectrophoretic structuring on piezoelectric and pyroelectric properties of PT-epoxy composites. Smart Mater. Struct. **23**, 105030 (2014)
11. V.G. Smotrakov, V.V. Eremkin, V.A. Aleshin, E.S. Tsikhotsky, Manufacturing and study of a single crystal—ceramic composite. Izvestiya RAN. Seriya Fizicheskaya **64**, 1220–1223 (2000) (in Russian)
12. H. Takahashi, S. Tukamoto, J. Qiu, J. Tani, T. Sukigara, Property of composite ceramics composed of single crystals and ceramic matrix using hybrid sintering. Jpn. J. Appl. Phys., Part 1 **42**, 6055–6058 (2003)
13. A. Chaipanich, Dielectric and piezoelectric properties of PZT–cement composites. Curr. Appl. Phys. **7**, 537–539 (2007)
14. S.E. Filippov, A.A. Vorontsov, V.Yu. Topolov, O.E. Brill, P. Bisegna, A.E. Panich, Features of the piezoelectric effect in a novel PZT-type ceramic/clay composite. Ferroelectr. Lett. Sect. **41**, 82–88 (2014)
15. P. Zhao, S. Kim, B. Hinderliter, Investigation of cement–sand-based piezoelectric composites. J. Intell. Mater. Syst. Struct. **27**, 1666–1672 (2016)
16. A. Safari, M. Allahverdi, E.K. Akdogan, Solid freeform fabrication of piezoelectric sensors and actuators. J. Mater. Sci. **41**, 177–198 (2006)
17. J.E. Smay, J. Cesarano III, A.B. Tuttle, J.A. Lewis, Piezoelectric properties of 3-X periodic $Pb(Zr_xTi_{1-x})O_3$—polymer composites. J. Appl. Phys. **92**, 6119–6127 (2002)
18. J.E. Smay, B. Tuttle, J. Cesarano III, Robocasting of three-dimensional piezoelectric structures, in *Piezoelectric and Acoustic Materials for Transducer Applications*, ed. by A. Safari, E.K. Akdoğan (Springer, New York, 2008), pp. 305–318
19. D. van den Ende, B. Bory, W. Groen, S. van der Zwaag, Improving the d_{33} and g_{33} properties of 0–3 piezoelectric composites by dielectrophoresis. J. Appl. Phys. **107**, 024107 (2010)
20. H. Khanbareh, *Expanding the Functionality of Piezo-Particulate Composites*. Dissertation, Delft University of Technology, Delft (2016)
21. D.M. Gureev, R.B. Ruzhechko, I.V. Shishlovsky, Selective laser sintering of PZT ceramic powders. Tech. Phys. Lett. **26**, 262–264 (2000)
22. I.V. Shishkovsky, *Laser Synthesis of Functionally Gradient Mesostructures and Volume Samples* (Fizmatlit, Moscow, 2009) (in Russian)
23. S. Sherrit, H.D. Wiederick, B.K. Mukherjee, PcC3 0–3 piezoelectric–glass composites. Ferroelectrics **134**, 65–69 (1992)
24. K. Tajima, H.J. Hwang, M. Sando, K. Niihara, PZT nanocomposites reinforced by small amount of oxides. J. Eur. Ceram. Soc. **19**, 1179–1182 (1999)
25. T.-Y. Chen, S.-Y. Chu, C.-K. Cheng, Doping effect of SiO_2 on the dielectric and piezoelectric properties of $PbTiO_3$-based ceramics and its SAW applications. Integr. Ferroelectr. **69**, 65–72 (2005)
26. C.R. Bowen, V.Yu. Topolov, H.A. Kim, *Modern Piezoelectric Energy-Harvesting Materials* (Springer International Publishing, Switzerland, 2016)
27. F. Levassort, M. Lethiecq, C. Millar, L. Pourcelot, Modeling of highly loaded 0–3 piezoelectric composites using a matrix method. IEEE Trans. Ultrason. Ferroelectr. Freq. Control **45**, 1497–1505 (1998)
28. L.V. Gibiansky, S. Torquato, On the use of homogenization theory to design optimal piezocomposites for hydrophone applications. J. Mech. Phys. Solids **45**, 689–708 (1997)
29. A.A. Grekov, S.O. Kramarov, A.A. Kuprienko, Effective properties of a transversely isotropic piezoelectric composite with cylindrical inclusions. Mech. Compos. Mater. **25**, 54–61 (1989)
30. K.E. Evans, K.L. Alderson, The static and dynamic moduli of auxetic microporous polyethylene. J. Mater. Sci. Lett. **11**, 1721–1724 (1992)
31. H.L.W. Chan, J. Unsworth, Simple model for piezoelectric ceramic/polymer 1–3 composites used in ultrasonic transducer applications. IEEE Trans. Ultrason. Ferroelectr. Freq. Control **36**, 434–441 (1989)
32. V.Yu. Topolov, P. Bisegna, C.R. Bowen, *Piezo-active Composites. Orientation Effects and Anisotropy Factors* (Springer, Berlin, Heidelberg, 2014)

33. T. Yamamoto, K. Urabe, H. Banno, $BaTiO_3$ particle-size dependence of ferroelectricity in $BaTiO_3$/polymer composites. Jpn. J. Appl. Phys., Part 1 **32**, 4272–4276 (1993)
34. T. Yamada, T. Ueda, T. Kitayama, Piezoelectricity of a high-content lead zirconate titanate/polymer composite. J. Appl. Phys. **53**, 4328–4332 (1982)
35. S.T. Lau, K.W. Kwok, F.G. Shin, S. Kopf, A poling study of lead zirconate titanate/ polyurethane 0–3 composites. J. Appl. Phys. **102**, 044104 (2007)
36. C.J. Dias, D.K. Das-Gupta, Poling behaviour of ceramic/polymer ferroelectric composites. Ferroelectrics **157**, 405–410 (1994)
37. Y.T. Or, C.K. Wong, B. Ploss, F.G. Shin, Modeling of poling, piezoelectric, and pyroelectric properties of ferroelectric 0–3 composites. J. Appl. Phys. **94**, 3319–3325 (2003)
38. X. Renxin, C. Wen, Z. Jing, L. Yueming, S. Huajun, Dielectric and piezoelectric properties of 0–3 PZT-PVDF composite doped with polyaniline. J. Wuhan Univ. Technol., Mater. Sci. Ed. **21**, 84–87 (2006)
39. G. de Campos Fuzari Jr., M.O. Orlandi, E. Longo, W.L. de Barros Melo, W.K. Sakamoto, Effect of controlled conductivity on thermal sensing property of 0–3 pyroelectric composite. Smart Mater. Struct. **22**, 025015 (2013)
40. W. Sakamoto, P. Marin-Franch, D. Das-Gupta, Characterization and application of PZT/PU and graphite doped PZT/PU composite. Sens. Actuators, A **100**, 165–174 (2002)
41. W.K. Sakamoto, G. de Campos Fuzari Jr, M.A. Zaghete, R.L.B. de Freitas, Lead titanate-based nanocomposite: fabrication, characterization and application and energy conversion evaluation. Ferroelectr. Mater. Asp., 251–276 (2011). https://www.intechopen.com/books/ferroelectrics-material-aspects
42. K.H. Chau, Y.W. Wong, F.G. Shin, Enhancement of piezoelectric and pyroelectric properties of composite films using polymer electrolyte matrix. Appl. Phys. Lett. **91**, 252910 (2007)
43. C. Wong, F. Shin, Effect of electrical conductivity on poling and the dielectric, pyroelectric and piezoelectric properties of ferroelectric 0–3 composites. J. Mater. Sci. **41**, 229–249 (2006)
44. G. Sa-Gong, A. Safari, S.J. Jang, R.E. Newnham, Poling flexible piezoelectric composites. Ferroelectr. Lett. Sect. **5**, 131–142 (1986)
45. W.K. Sakamoto, E. de Souza, D.K. Das-Gupta, Electroactive properties of flexible piezoelectric composites. Mater. Res. **4**, 201–204 (2001)
46. X.F. Liu, C.X. Xiong, H.J. Sun, L.J. Dong, R. Li, Y. Liu, Piezoelectric and dielectric properties of PZT-PVC and graphite doped with PZT-PVC composites. Mater. Sci. Eng., B **127**, 261–266 (2006)
47. B. Noheda, D.E. Cox, G. Shirane, R. Guo, B. Jones, L.E. Cross, Stability of the monoclinic phase in the ferroelectric perovskite $PbZr_{1-x}Ti_xO_3$. Phys. Rev. B **63**, 014103 (2001)
48. L.A. Reznitchenko, L.A. Shilkina, O.N. Razumovskaya, E.A. Yaroslavtseva, S.I. Dudkina, O.A. Demchenko, YuI Yurasov, A.A. Esis, I.N. Andryushina, Phase x–T diagram of actual solid solutions of the (1–x)$PbZrO_3$ – x$PbTiO_3$ system ($0.37 \leq x \leq 0.57$). Phys. Solid State **50**, 1527–1533 (2008)
49. N. Zhang, H. Yokota, A.M. Glazer, Z. Ren, D.A. Keen, D.S. Keeble, P.A. Thomas, Z.-G. Ye, The missing boundary in the phase diagram of $PbZr_{1-x}Ti_xO_3$. Nat. Commun. **5**, 5231 (2014)
50. D.A. van den Ende, Structured piezoelectric composites, materials and applications, Ph. D. thesis, Faculty of Aerospace Engineering, Delft University of Technology, Delft (2012)
51. C. Dias, D. Das Gupta, Inorganic ceramic polymer ferroelectric composite electrets. IEEE Trans. Dielectr. Electr. Insul. **3**, 706–734 (1996)
52. H. Khanbareh, Groen W.A. van der Zwaag, In-situ poling and structurization of piezoelectric particulate composites. J. Intell. Mater. Syst. Struct. **28**, 2467–2472 (2017)
53. C.P. Bowen, R.E. Newnham, C.A. Randall, Dielectric properties of dielectrophoretically assembled particulate-polymer composites. J. Mater. Res. **13**(1998), 205–210 (1998)
54. H. Tang, Y. Lin, H.A. Sodano, Synthesis of high aspect ratio $BaTiO_3$ nanowires for high energy density nanocomposite capacitors. Adv. Energy Mater. **2**, 451–456 (2012)
55. H. Tang, M.H. Malakooti, H.A. Sodano, Relationship between orientation factor of lead zirconate titanate nanowires and dielectric permittivity of nanocomposites. Appl. Phys. Lett. **103**, 222901 (2013)

56. K.W. Kwok, S.T. Lau, C.K. Wong, F.G. Shin, Effect of electrical conductivity on poling of ferroelectric composites. J. Phys. D Appl. Phys. **40**, 6818–6823 (2007)
57. T.B. Jones, *Electromechanics of Particles* (Cambridge University Press, Cambridge, New York, 2005)
58. L. Li, S. Zhang, Z. Xu, F. Wen, X. Geng, H.J. Lee, T.R. Shrout, 1–3 piezoelectric composites for high-temperature transducer applications. J. Phys. D: Appl. Phys. **46**, 165306 (2013)
59. J. Lee, B. Choi, Development of a piezoelectric energy harvesting system for implementing wireless sensors on the tires. Energy Convers. Manag. **78**, 32–38 (2014)
60. Q. Xue, M. Stanton, G. Elfbaum, A high temperature and broadband immersion 1–3 piezocomposite transducer for accurate inspection in harsh environments. Proc. IEEE Ultrason. Symp. **2**, 1372–1375 (2003)
61. H.J. Lee, S. Zhang, Y. Bar-Cohen, S. Sherrit, High temperature, high power piezoelectric composite transducers. Sensors **14**, 14526–14552 (2014)
62. C. Park, Z. Ounaies, K.E. Wise, J.S. Harrison, In situ poling and imidization of amorphous piezoelectric polyimides. Polymer **45**, 5417–5425 (2004)
63. Z. Ounaies, C. Park, J.S. Harrison, J.G. Smith, J. Hinkley, Structure—property study of piezoelectricity in polyimides. Proc. SPIE **3669**, 171–178 (1999)
64. G.M. Atkinson, R.E. Pearson, Z. Ounaies, C. Park, J.S. Harrison, S. Dogan, J.A. Midkiff, Novel piezoelectric polyimide MEMS, in *Transducers 2003—12th International Conference on Solid-State Sensors, Actuators and Microsystems, Digest of Technical Papers*, vol. 1 (2003), pp. 782–785
65. J. Simpson, Z. Ounaies, C. Fay, Polarization and piezoelectric properties of a nitrile substituted polyimide. Mater. Res. Soc. Symp.—Proc. **459**, 59–64 (1997)
66. Z. Ounaies, C. Park, J. Harrison, P. Lillehei, Evidence of piezoelectricity in SWNT-polyimide and SWNT-PZT-polyimide composites. J. Thermoplast. Compos. Mater. **21**, 393–409 (2008)
67. B. Jaffe, R.S. Roth, S. Marzullo, Piezoelectric properties of lead zirconate-lead titanate solid-solution ceramics. J. Appl. Phys. **25**, 809–810 (1954)
68. M. Lines, A. Glass, *Principles and Application of Ferroelectrics and Related Materials* (Clarendon Press, Oxford, 1977)
69. G.A. Smolensky, V.A. Bokov, V.A. Isupov, N.N. Krainik, R.E. Pasynkov, A.I. Sokolov, N.K. Yushin, *Physics of Ferroelectric Phenomena* (Nauka, Leningrad, 1985) (in Russian)
70. Y. Xu, *Ferroelectric Materials and Their Applications* (North-Holland, Amsterdam London New York Toronto, 1991)
71. A.V. Turik, V.Yu. Topolov, A.I. Chernobabov, Domain switching and anisotropy of piezoelectric moduli in $PbTiO_3$-type ceramics. Electroceramics IV—4th International Conference on Electronic Ceramics & Applications. September 5–7, 1994, Aachen, Germany. Proceedings, vol. 1 Dielectric and microwave materials, ferroelectrics, piezoelectrics, relaxors, grain boundary controlled materials. Aachen (1994), pp. 519–522
72. A.V. Turik, V.Yu. Topolov, Ferroelectric ceramics with a large piezoelectric anisotropy. J. Phys. D Appl. Phys. **30**, 1541–1549 (1997)
73. K.C. Khemani, *Polymeric Foams* (American Chemical Society, Washington, 1997)
74. H. Khanbareh, K. de Boom, S. van der Zwaag, W.A. Groen, Highly sensitive piezo particulate-polymer foam composites for robotic skin application. Ferroelectrics **515**, 25–33 (2017)
75. D. Krause, G.-H. Koops, N.F. van der Vegt, M. Wessling, M. Wubbenhorst, J. van Turnhout, Ultralow-k dielectrics made by supercritical foaming of thin polymer films. Adv. Mater. **14**, 1041–1046 (2002)
76. C. Dias, Ferroelectric composites for pyro- and piezoelectric applications. Ph.D. thesis, School of Electronic Engineering and Computer Systems, University of Wales, Bangor (1994)
77. N. Chand, J. Sharma, Influence of porosity on resistivity of polypropylene foams. J. Cell. Plast. **48**, 43–52 (2012)

Chapter 3
Experimental Studies on Effective Properties and Related Parameters of Piezo-Particulate Composites

Abstract Numerous experimental results on the effective dielectric, pyroelectric and piezoelectric properties and related parameters of piezo-particulate composites based on ferroelectric ceramics are described and analysed. The effective properties of the composites structured by means of dielectrophoresis (0–3 and/or 1–3 connectivity patterns) are compared to the properties of related random (or non-structured) composites with 0–3 connectivity. The influence of the composite microgeometry and properties of the components on the effective properties is discussed. Large values of maximum of the piezoelectric coefficient g_{33}^* describing piezoelectric sensitivity are achieved in the structured composite. The considerable increase of g_{33}^* is a result of forming a porous structure in the polymer matrix of the composite. Examples of the influence of the porous (foam) polymer matrix on the piezoelectric coefficient g_{33}^* are considered for composite based on the ferroelectric PZT ceramic. High-temperature piezo-active composites exhibit thermal stability of both the dielectric and piezoelectric properties in the presence of the ferroelectric ceramic component with the high Curie temperature T_C and due to the considerable thermal stability of the dielectric properties of the polymer matrix.

An interest in the study on the effective physical properties of the piezo-active composites [1–6] originates mainly from a link between their microgeometry and properties [7]. The link becomes more complex in the presence of a system of randomly distributed inclusions or a range of piezoelectric components in the composite. Attractive combinations of mechanical, thermal, dielectric, acoustic, piezoelectric and other properties and ease of processability can be achieved by embedding a large number of FC particles (inclusions) in a polymer matrix. Such a manufacturing operation leads to the formation of a composite with 0–3 connectivity in terms of work by Newnham et al. [8]. In the 0–3 FC/polymer composite, ferroelectric activity and related properties (e.g. piezoelectric and pyroelectric) is achieved due to the poled FC component, and mechanical flexibility is provided by the compliant polymer component. The 0–3 FC/polymer composite configuration can overcome some limitations of both the FC and polymer components and, therefore, becomes an important active element of many piezoelectric transducers, sensors, energy harvesting and other devices.

H. Khanbareh et al., *Piezo-Particulate Composites*, Springer Series in Materials Science 283, https://doi.org/10.1007/978-3-030-19204-4_3

In the present chapter, we describe experimental results on the effective physical properties and related parameters of the piezo-particulate composites based on FCs. We also discuss ways to improve some parameters of these composites for specific piezo-technical applications.

3.1 Microgeometry of Piezo-Particulate Composites

The term 'microgeometry' is often used along with the term 'connectivity' to characterise the specifics of the spatial distribution of components (or structural elements) in the composite sample. Interconnections between the composite microgeometry and specific connectivity pattern in the composite sample often facilitate a prediction of the effective properties for such connectivities as 1–3, 0–3, 3–3, etc. Figure 3.1 shows that knowledge of the connectivity and microgeometry is not sufficient to describe the composite architecture in detail. The next important term that makes this description easier is 'microstructure'. The term 'microstructure' is often used to describe the characterisation of the granular structure of FCs, air pores or channels and defects therein [9–13]. In the case of a composite, its microstructure is to be studied in the context of the microstructures of each component. The complex microstructure of many FCs can be an additional obstacle at the description of the composite and the prediction of its effective physical properties. Taking into account the ferroelectric nature of the main component of the composite, we should know specifics of the domain structure and its changes under external electric and/or mechanical fields.

In describing the microgeometry and microstructural features of a two-component composite, we can draw an analogy with stones and rocks found in nature. In Fig. 3.2, we show some examples and underline that a similar microgeometry does not necessarily mean a similar microstructure. Various types of inclusions (or rods, layers, and so on) as well as their distribution over the geological formation can be described in terms of the connectivity patterns such as 0–3 (Fig. 3.2a, b), 2–2 (Fig. 3.2c), 0–3 and 3–3 (Fig. 3.2d). The 'architecture' of several stones becomes highly complex when many inclusions are heterogeneous and composite-like; see, for instance, Fig. 3.2d. The composite-like 'architecture' of the rock in Fig. 3.2e suggests that the 1–3, 2–2 and 0–3 connectivity patterns can be taken into account in their characterisation. Finally, in Fig. 3.2f, g, we show the rock (profile and face, respectively) with the well-developed composite-like 'architecture', and the main connectivity patterns suitable for the description are of 2–2 and 0–3 types. Undoubtedly, features of the microstructure of the stones or blocks which form the rocks (Fig. 3.2e–g) are more various than features of the individual samples shown in Fig. 3.2a–d.

In Fig. 3.3a–c, we show examples of the non-structured and structured piezo-active composites wherein the PT FC particles are distributed in the polymer medium. Hereby an analogy with the 'fish ensemble' in sea water (Fig. 3.3d, e) may be taken into consideration at the description of the structuring process in the 0–3-type composite. Orientations of the rectangles that surround the fish in Fig. 3.3e suggest that the majority of the fish would be oriented almost along the vertical line.

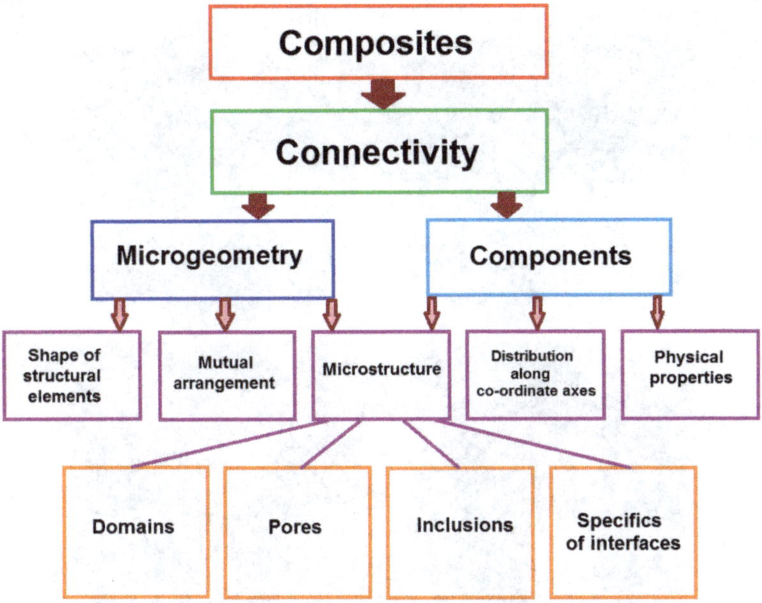

Fig. 3.1 Links between microgeometric characteristics of composites

3.2 Ferroelectric Behaviour and Related Parameters

Figures 3.4 and 3.5 show examples of the ferroelectric behaviour of the non-structured (or random) and DEP-structured FC-based composites with 0–3 connectivity patterns. The composite samples are studied by the experimental methods described in Chap. 2 and work [14, 15]. A comparison of experimental data on the non-structured and structured composites enables us to state the following. Increasing the volume fraction of FC m leads to an increase of the remanent polarisation of the composite P_r^* and to an almost constant coercive field $E_c = 9$ kV/mm, see Fig. 3.4. This E_c value is approximately seven times larger than the coercive field of the bulk PZT5A4 FC (see Sect. 2.4.2) which is the piezoelectric component of the studied composites. At a volume fraction of FC m = const from the studied range, the structured composite exhibits larger values of the electric displacement D_3 and, therefore, the polarisation P_3 compared to the non-structured composite, see Fig. 3.4. The largest value of the remanent polarisation $P_r^* = 0.18$ C/m^2 is achieved in the structured composite at $m = 0.50$. This is accounted by the active role of the FC component in forming the ferroelectric properties of the composite and by the favourable arrangement of the prolate FC inclusions along the $\boldsymbol{E}$ vector. Comparing the graphs in Fig. 3.4, we state that the dielectric permittivity $\varepsilon_{33}^{*\sigma} = dD_3/dE_3 = \Delta D_3/\Delta E_3$ at the electric field $E_3 \to 0$ increases on increasing the volume fraction of FC m. At the electric field $E_3 >$ 18 kV/mm, an electrical breakdown of the studied composite samples is observed.

Fig. 3.2 Stones (**a**–**d**) and rock formations (**e** and **f**) as examples of natural composites. The stones (**a**–**c**) have been collected on the Black sea coast, the stone (**d**) has been collected on the Azov sea coast, and the rocks have been taken near Anapa (**e**) and Goryachy Klyuch (**f** and **g**), Krasnodar Territory (Russia)

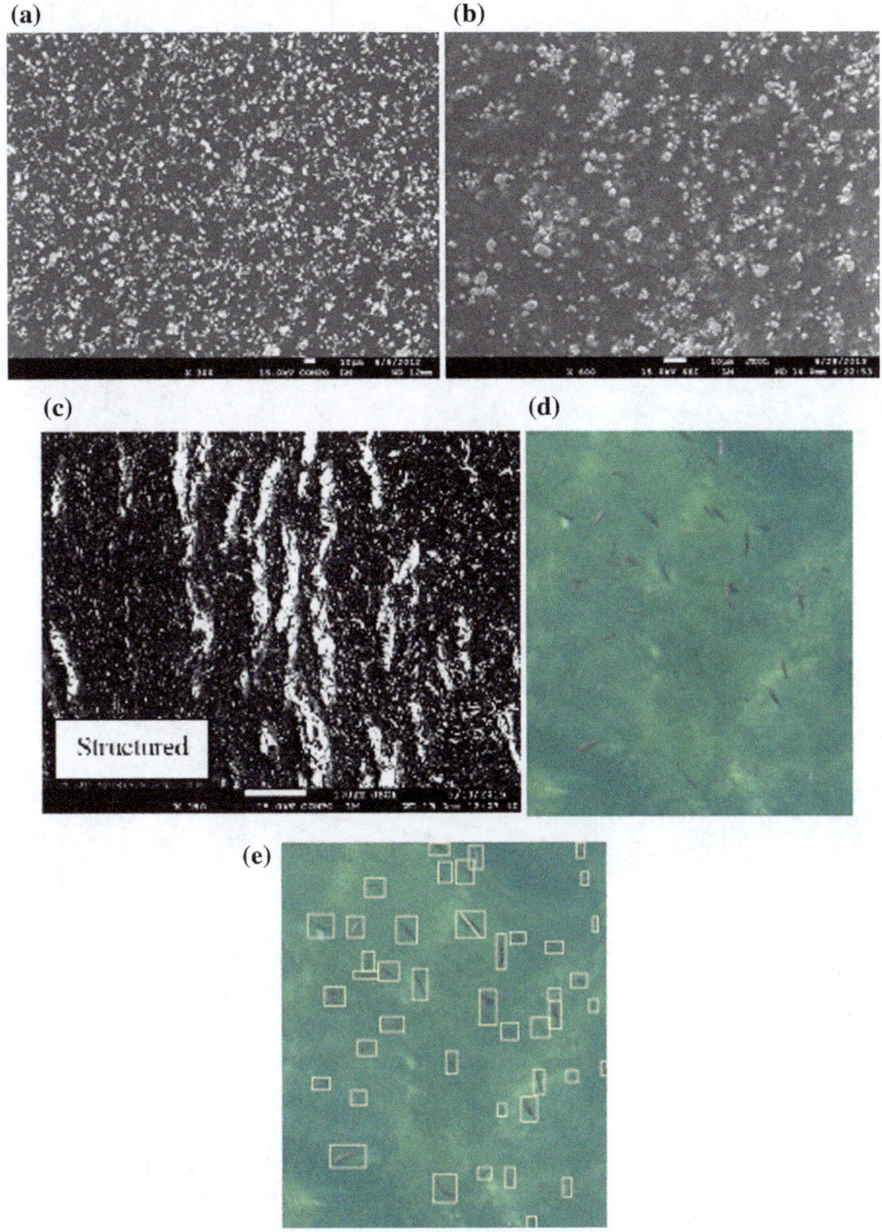

Fig. 3.3 Micrographs of the non-structured (**a** and **b**) and structured (**c**) PT FC/polymer composites and 'fish ensemble' in Black sea water (Russia) as an analogue of the structured composite (**d** and **e**). An arrangement within the 'fish ensemble' is shown by means of rectangles (**e**). DEP has been applied to obtain the structured composite (**c**)

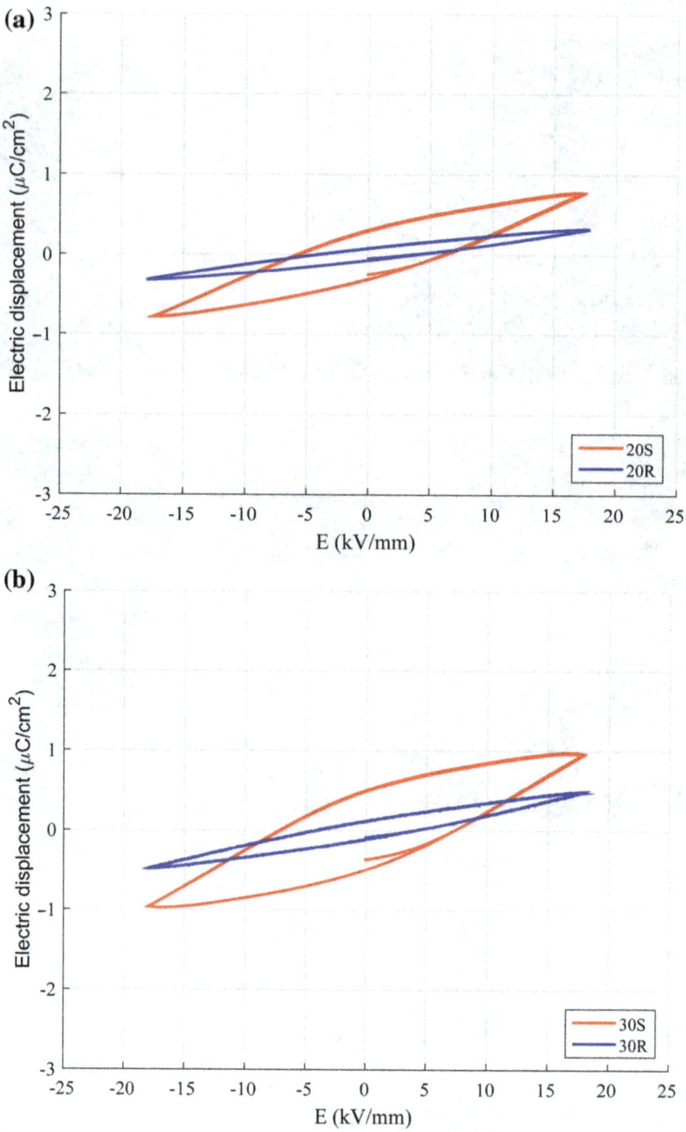

Fig. 3.4 Electric displacement—electric field ($D_3 - E_3$) loop built for the PZT5A4 FC/epoxy composite at $m = 0.2$ (**a**), 0.3 (**b**), 0.4 (**c**), and 0.5 (**d**). R and S are related to the random and structured composites, respectively [15]

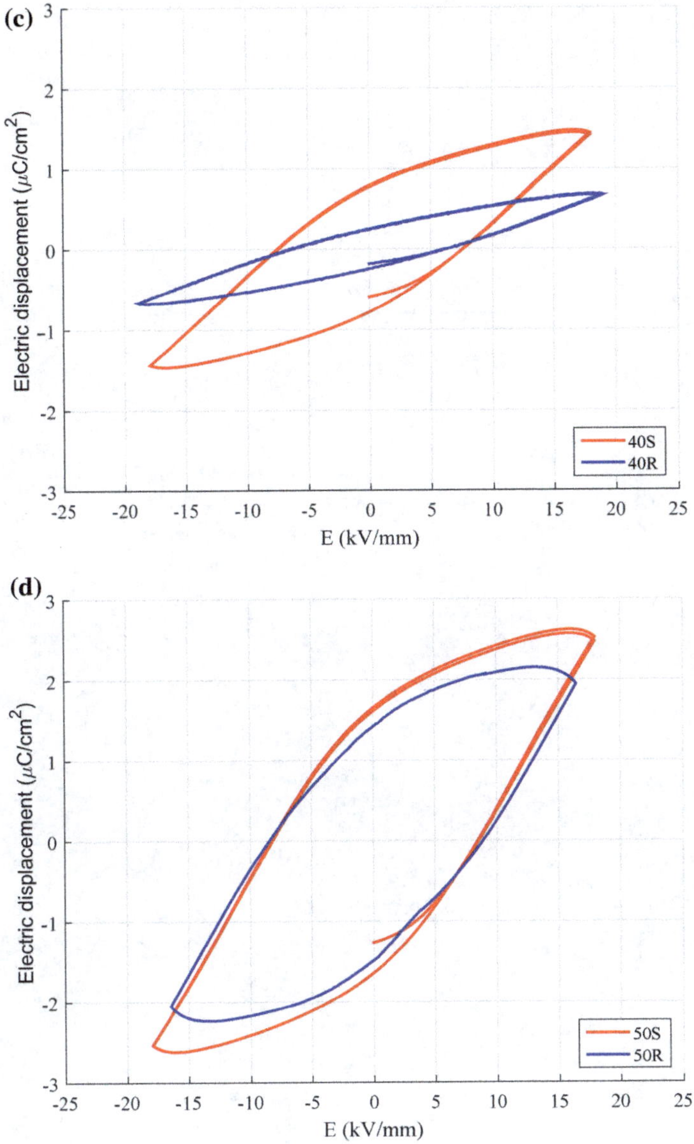

Fig. 3.4 (continued)

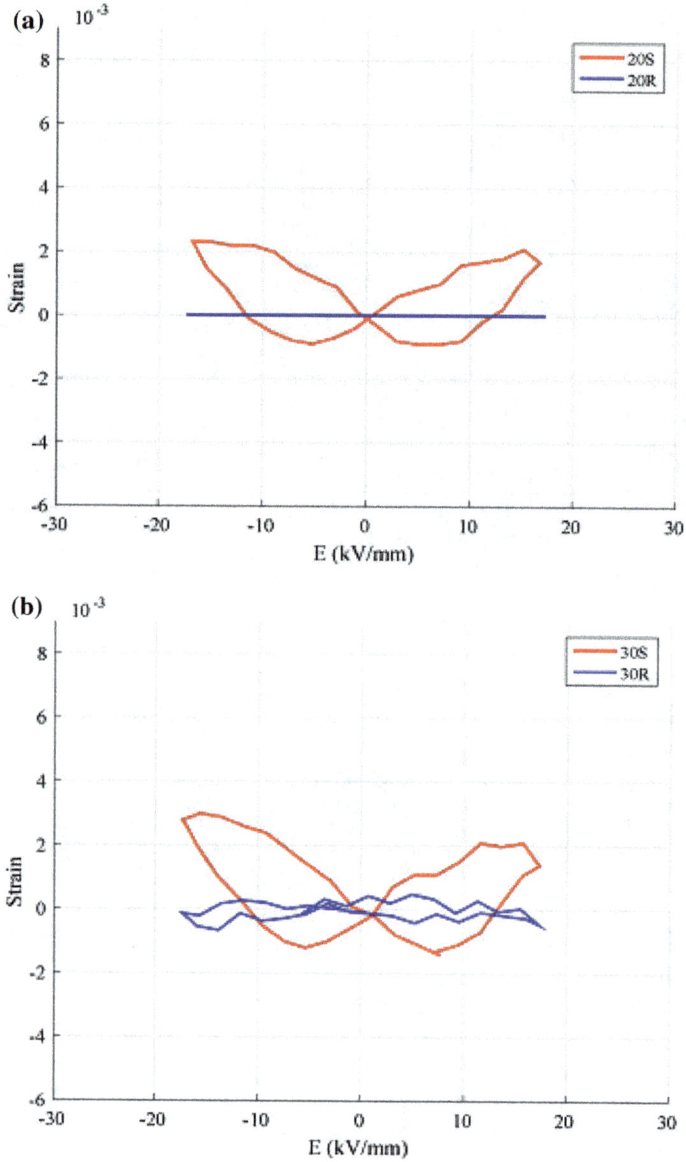

Fig. 3.5 Mechanical strain—electric field ($\xi_3 - E_3$) loop built for the PZT5A4 FC/epoxy composite at $m = 0.2$ (**a**), 0.3 (**b**), 0.4 (**c**), and 0.5 (**d**). *R* and *S* are related to the random and structured composites, respectively [15]

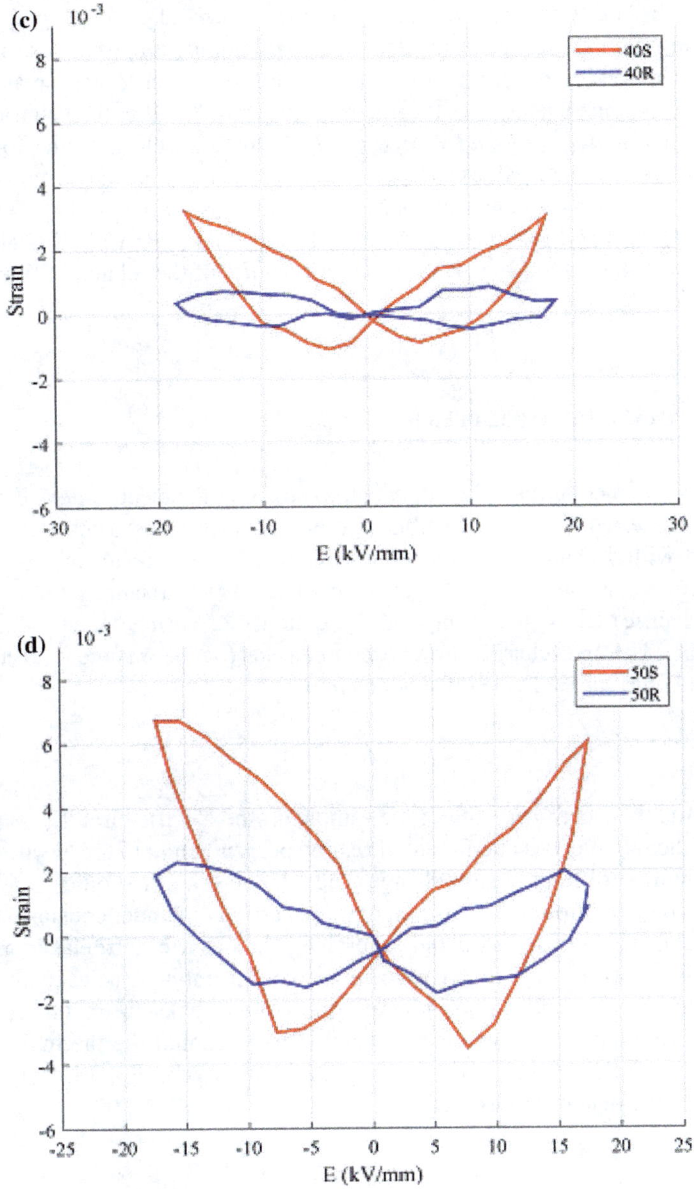

Fig. 3.5 (continued)

Changes in the 'butterfly' configuration of the curves shown in Fig. 3.5 suggest that the longitudinal piezoelectric effect in the structured composite is stronger than the non-structured composite at a volume fraction m = const, on increasing m, an increase of the piezoelectric coefficient d_{33}^* is observed, and at the electric field $E_3 \to 0$, the relation $d_{33}^* = d\xi_3/dE_3 = \Delta\xi_3/\Delta E_3$ holds. As follows from experimental data shown in Fig. 3.5, values of $d_{33}^* \sim 10^2$ pC/N are achieved in the studied m range, and the order-of-magnitude of d_{33}^* is the same as for the poled PZT5A4 FC.

It should be noted that the d_{33}^* value evaluated for the PZT5A4 FC at $E_3 \to 0$ (see experimental data in Fig. 2.3b) is larger than d_{33}^* of the related DEP-structured composite.

3.3 Pyroelectric Properties

Due to the presence of the FC component poled along the electric field $\boldsymbol{E}$ direction, the DEP-structured composite exhibits pyroelectric properties. These properties are associated with a change of spontaneous polarisation as a result of a temperature change [16]. As is known, the change in spontaneous polarisation of a pyroelectric (or ferroelectric) leads to a change in the quantity of surface electric charges on this sample. The free charges flow to compensate for the change, and leads to a pyroelectric current i_p [16–18] that is given by

$$i_p = Ap\,\mathrm{d}T/\mathrm{d}t. \tag{3.1}$$

In (3.1), A is the electroded area of the sample on the electric circuit, p is the component of the pyroelectric coefficient measured perpendicular to the electrode plane, dT/dt is the rate of temperature change, and t is the time. It should be added that (3.1) holds on condition that measurements are carried out under constant stress and electric field. In this case, we avoid excessive piezoelectric, ferroelastic and ferroelectric contributions [16] into the pyroelectric coefficient p. To select a pyroelectric material for specific applications, several figures of merit have been introduced [19], and these parameters are calculated by taking into account the thermal, electrical and other properties of the material. The pyroelectric figure of merit that is used to characterise the material most frequently is

$$F = p/\varepsilon \tag{3.2}$$

where ε is the dielectric permittivity measured along the spontaneous polarisation vector. In the case of the piezo-active composites, we assume that the parameters from (3.2) obey conditions $\varepsilon = \varepsilon_{33}^{*\sigma}/\varepsilon_0$ and $p = p_3^{*\sigma}$.

The pyroelectric coefficient p is found from the reversible pyroelectric current i_p after three runs of the thermally stimulated discharge current measurement [14, 15], and (3.1) is used to calculate p for the random and DEP-structured composite samples.

A small change in the intensity of the discharge current was observed [15] during the first and second runs due to the activity of the space charges on the samples. The third heating cycle shows no reduction in the current, thus a reversible pyroelectric current was established. The effect of the DEP structuring on the pyroelectric properties of the PT/epoxy composite is illustrated with Fig. 3.6. The pyroelectric coefficient p from (3.2) increases as the volume fraction of PT m increases in the composite. Moreover, there is a mild improvement due to the DEP structuring of the composite samples [14, 15].

A model by Wong and Shin [20] can be applied [14, 15] to interpret the pyroelectric properties of a random FC/polymer composite. In this model, the composite is considered as a system of spherical FC particles that are uniformly dispersed in a continuous polymer medium wherein an electric conductivity affects an internal electric field in the composite sample and, therefore, an accumulation of charges at the FC–polymer interfaces appears therein. Results of the prediction of the volume-fraction behaviour of the pyroelectric coefficient p are shown in Fig. 3.6a, and these results are consistent with the experimental results related to the random PT/epoxy composite [14, 15]. It should be noted that the model from work [20] does not take into consideration the features of the connectivity of the composite.

The pyroelectric figure of merit F from (3.2) increases (Fig. 3.6b) as the volume fraction of PT increases and improvement of F is observed due to the DEP structuring. This means that improved 'agreement' between the pyroelectric coefficient p and dielectric permittivity ε of the composite sample is achieved in the presence of the poled chain-like FC inclusions. In Fig. 3.7, an example of the temperature behaviour of the pyroelectric coefficient $p(T)$ is shown for the same PT-based composite. We observe a considerable increase in $p(T)$ at $T > 60$ °C. This transition temperature is in well agreement with the measured glass transition temperature T_g of the polymer matrix [14, 15]. The increase in $p(T)$ may be due to the fact that at $T > T_g$ of the polymer matrix, an increase of the free volume charge contributing to a high current [21] takes place. It should be added that the similar $p(T)$ dependence is typical of many FC/polymer composites, see work [21–25].

The volume-fraction dependence of the pyroelectric coefficient p of the composite at $T = 25$ °C and $T = 70$ °C is shown for comparison in Fig. 3.8. The considerable improvement of the pyroelectric activity of the composite is observed at $T = 70$ °C. It should be noted that in the vicinity of $T = 70$ °C, the largest dp/dT value is achieved (Fig. 3.7) for both the non-structured (random) and structured composites based on PT.

The next example of the pyroelectric properties is concerned with a composite where a polyethylene oxide (PEO) matrix contains PT inclusions [14]. We remind the reader that PEO is a synthetic polymer that is often used in the pure non-ionic state as a surfactant and as a water-soluble thermoplastic resin as well as an ionic conducting polymer for Li-batteries, displays, sensors and other electrochemical devices [26]. PEO is characterised by good structural integrity and low glass transition temperature T_g. The PT/PEO composite is poled in the electric field $E = 10$ kV/mm for 15 min [14]. The high pyroelectric performance in the studied volume-fraction range (Fig. 3.9) is achieved due to the higher electric conductivity of the PEO matrix. In the

(a)

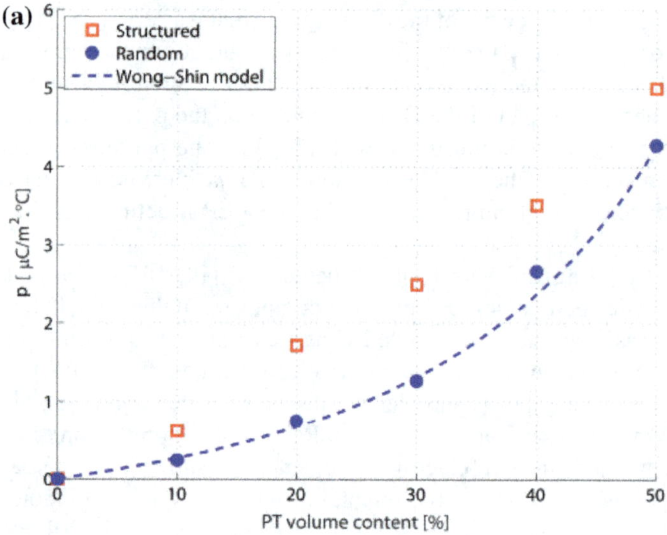

(b)

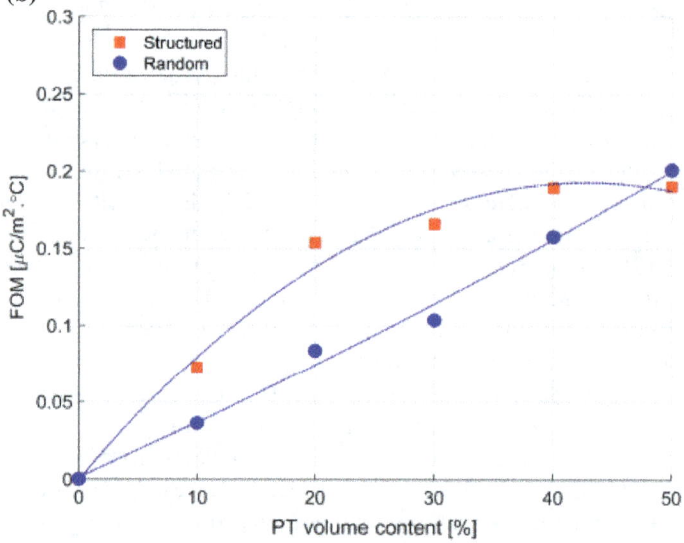

Fig. 3.6 Pyroelectric coefficient p [a, in 10^{-6} C/(m^{2}K)] (reprinted from paper by Khanbareh et al. [14], with permission from IOP Publishing) and pyroelectric figure of merit F [b, in 10^{-6} C/(m^{2}K)] of the PT/epoxy composite [15]

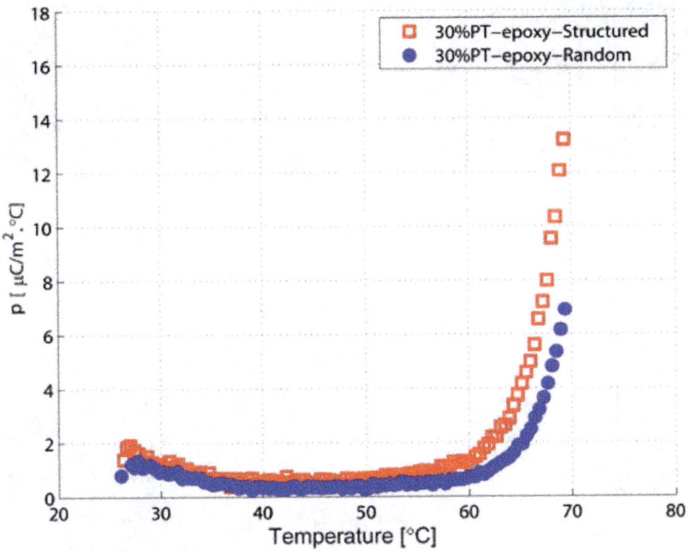

Fig. 3.7 Temperature dependence of the pyroelectric coefficient p [in 10^{-6} C/(m^2 K)] of the PT/epoxy composite at the volume fraction of PT FC $m = 0.3$ (reprinted from paper by Khanbareh et al. [14], with permission from IOP Publishing)

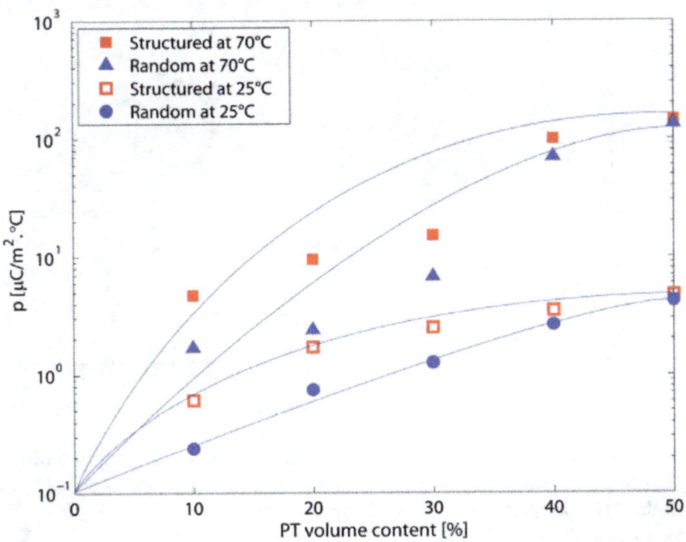

Fig. 3.8 Pyroelectric coefficient p [in 10^{-6} C/(m^2 K)] of the PT/epoxy composite manufactured at $T = \text{const}$ (reprinted from paper by Khanbareh et al. [14], with permission from IOP Publishing). Curves are built for eye guide

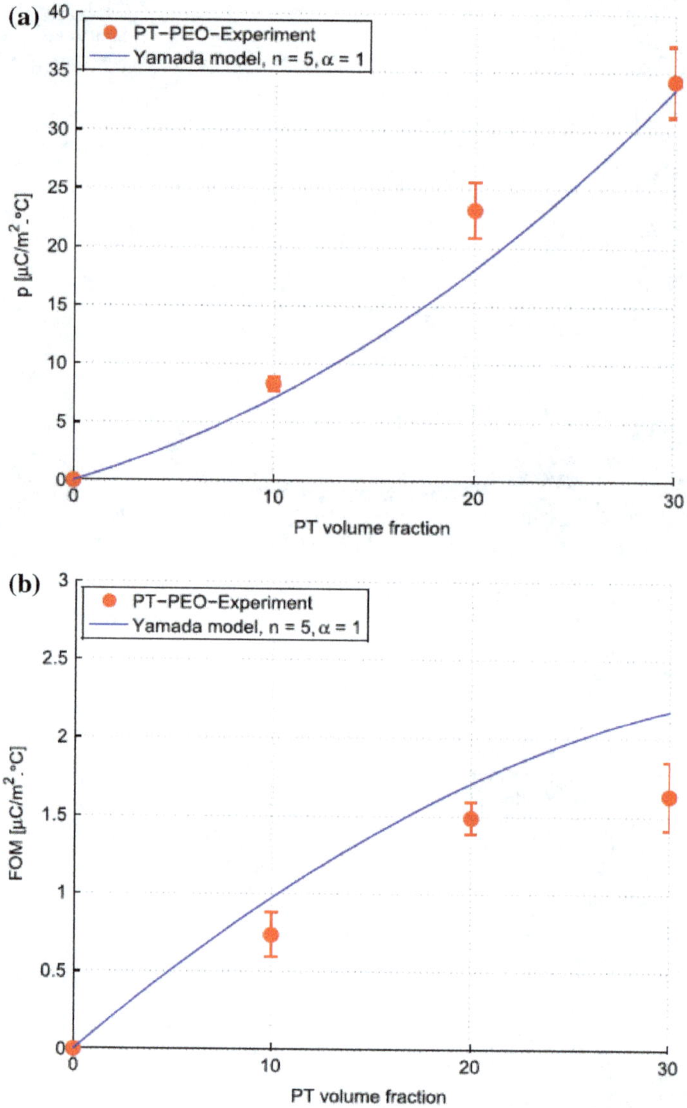

Fig. 3.9 Pyroelectric coefficient p [a, in 10^{-6} C/(m^2 K)] and pyroelectric figure of merit F [b, in 10^{-6} C/(m^2 K)] of the PT/PEO composite. Curves are built for eye guide (reprinted from paper by Khanbareh et al. [14], with permission from IOP Publishing)

presence of such a polymer matrix, the dielectric permittivity mismatch decreases by using a conducting matrix [14], and an electric flux path is formed between the FC particles. Due to these features of the properties, the local electric field acting on each FC particle is enhanced in the composite sample, and this facilitates the poling process and promotes improving the pyroelectric properties. The experimental results are also interpreted in terms of the composite model by Yamada et al. [27] where n and α are fitting parameters in this model.

3.4 Enhancing Piezoelectric Sensitivity

In addition to the pyroelectric properties for thermal sensing, the piezoelectric properties allow the material to be used as a mechanical sensor. The piezoelectric sensitivity of the FC/polymer composite is often characterised [5, 7] by the piezoelectric coefficients g_{3j}^* from (2.2). Figure 3.10 shows an example of the longitudinal piezoelectric sensitivity that is described by g_{33}^*. This piezoelectric coefficient is calculated by taking into account the experimental values of the piezoelectric coefficient d_{33}^* and dielectric permittivity $\varepsilon_{33}^{*\sigma}$ in the wide volume-fraction (m) range. Maxima of g_{33}^* are concerned with the effect of combination of properties [5] in the 0–3 and 1–3 composites based on FC, and the larger $\max g_{33}^*$ value is achieved in the DEP-structured composite mainly due to the larger piezoelectric coefficient d_{33}^* at m = const. It should be noted that the d_{33}^* increase is higher than the $\varepsilon_{33}^{*\sigma}$ increment with increasing the volume fraction of FC m in the DEP-structured composite, especially at $m < 0.1$. A location of $\max g_{33}^*$ (m) is dependent on the ratio of elastic moduli of the FC and polymer components [5, 14, 28, 29].

An important opportunity to increase the piezoelectric sensitivity and related piezoelectric coefficient g_{33}^* of the FC-based composite is concerned with forming a porous structure in its polymer matrix [5, 7, 30, 31]. In the porous 0–3-type composite, the piezoelectric coefficient d_{33}^* remains almost unchanged at a volume fraction of FC m = const, while the dielectric permittivity $\varepsilon_{33}^{*\sigma}$ decreases in comparison to $\varepsilon_{33}^{*\sigma}$ of the related composite with the monolithic polymer matrix. In work [30], the effective properties are reported for a PZT FC/porous polyurethane composite wherein a system of well-distributed spherical pores has been formed in the matrix (polyurethane foam).

An effect of D-water on microstructures of the polyurethane matrix is shown in Fig. 3.11. The microstructures shown in Fig. 3.11 suggest an increase of the gaseous volume fraction and the polymer sample thickness with increasing the volume fraction of D-water content. Porosity in the polyurethane matrix, and therefore, in the composite sample increases on increasing the volume fraction of D-water, see Fig. 3.12. The volume fraction of the open cells in the porous polyurethane samples is calculated by taking into account the area fraction of the open cells and the total area of the cells. As follows from results [15], the open-pore percentage monotonously increases from ca. 5 to 30% on increasing the volume fraction of D-water from 0 to

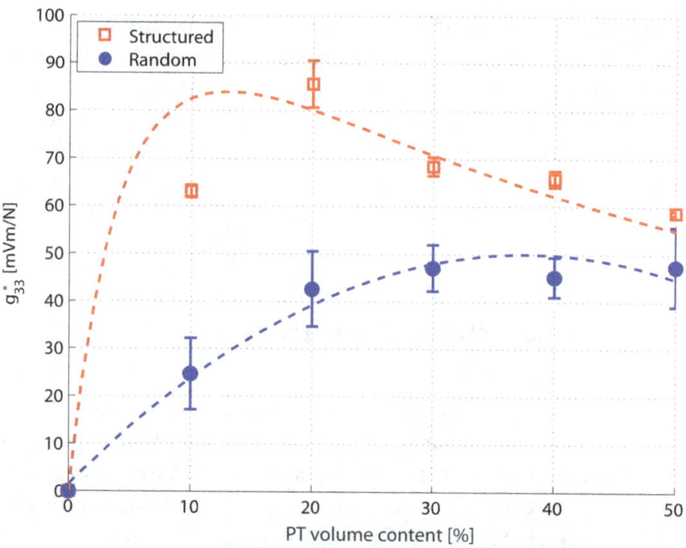

Fig. 3.10 Volume-fraction dependence of the piezoelectric coefficient g^*_{33} (in mV m/N) of the PT/epoxy composite [14] (reprinted from paper by Khanbareh et al. [14], with permission from IOP Publishing)

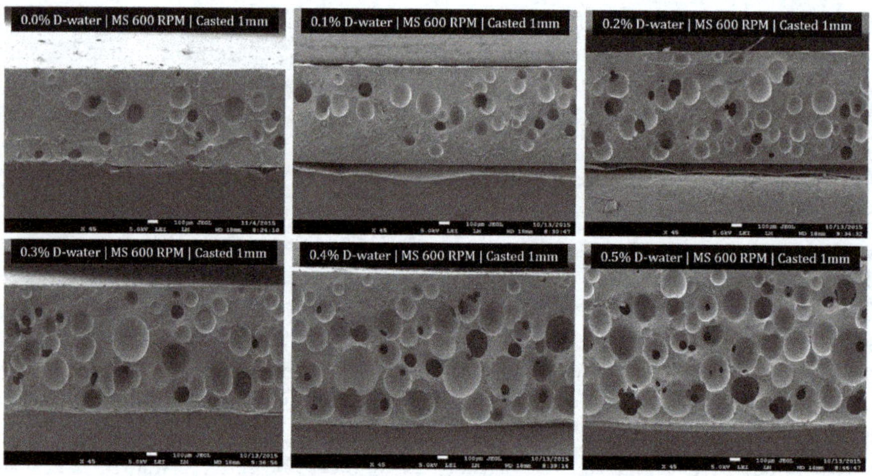

Fig. 3.11 SEM micrographs of polyurethane samples that contain different volume fractions of D-water, ranging from 0 to 0.5% (reprinted from paper by Khanbareh et al. [30], with permission from Taylor & Francis). The samples have been manufactured as described in work [15, 30], see also Sect. 2.7

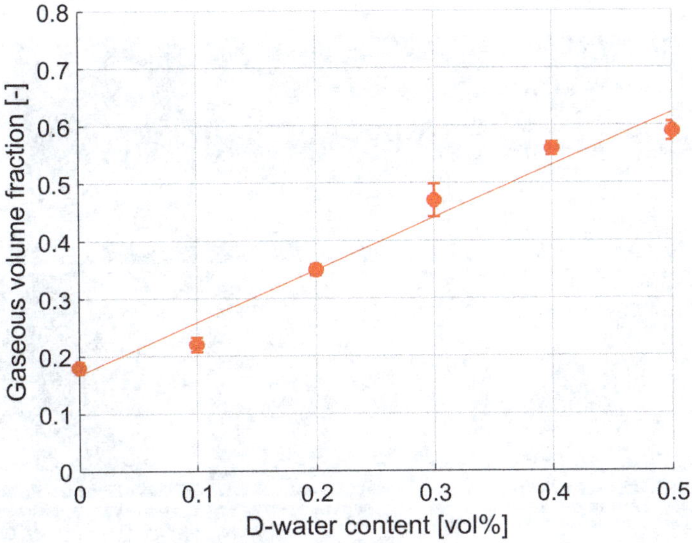

Fig. 3.12 Gaseous volume fraction versus D-water content in polyurethane samples (reprinted from paper by Khanbareh et al. [30], with permission from Taylor & Francis)

0.5%. At changes of the volume fractions of D-water in the same range, an average aspect ratio of the closed pores in the studied samples is approximately 1 [15].

Examples of typical microstructures of the random (or non-structured) and DEP-structured PZT FC/porous polyurethane composites are shown in Fig. 3.13a, b, respectively. The porous polyurethane matrix of the random composite has been prepared by using 0.4 vol% D-water. A similar matrix of the DEP-structured composite has been prepared using 0.2 vol% D-water. In the studied composite samples, an almost uniform distribution of FC particles is observed at volume fractions of FC $m < 0.4$, and at $m = 0.4$ agglomerations of the FC particles are observed. As follows from Fig. 3.13, a relatively uniform distribution of the pores through the thickness of the material is observed, and such a distribution is likely that affected by the application of the AC electric field during DEP. The FC chain formation is clearly visible in the cell walls, especially at a volume fraction $m = 0.1$. The DEP-structured composite sample exhibits a very rough surface due to the high degree of interface porosity [15, 30]. Elements of 1–3–0 connectivity are often observed in the DEP-structured samples, see Fig. 3.13b.

The influence of DEP structuring on the physical properties of the manufactured PZT-based composite is shown in Fig. 3.14. Hereby three composite models, that are traditionally applied to the 0–3 FC/polymer composites [27, 28, 32], can be used to interpret the obtained experimental results [15, 30]. The increase of the dielectric permittivity $\varepsilon_{33}^{*\sigma}$ of the DEP-structured composite in comparison to its analogue at $m = \text{const}$ (see Fig. 3.14a) is mainly due to the larger piezoelectric coefficient d_{33}^{*} (see Fig. 3.14b) and its contribution in the dielectric permittivity $\varepsilon_{33}^{*\sigma}$ in accordance

(a)

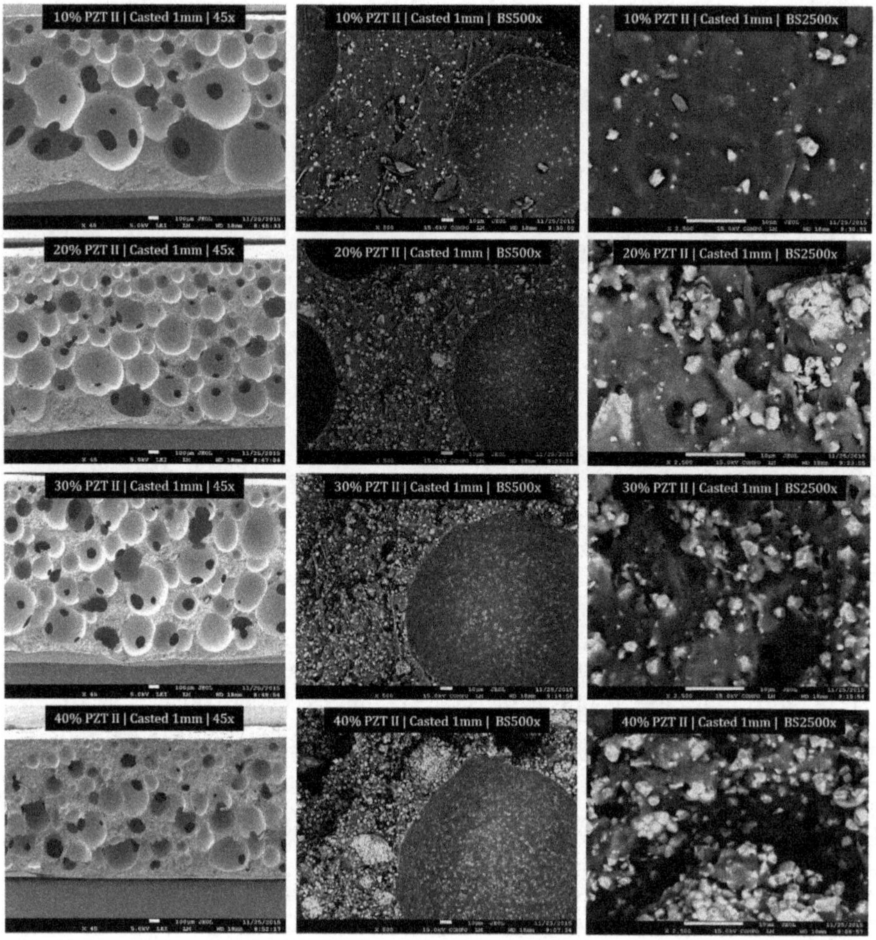

Fig. 3.13 SEM micrographs of the random (**a**) and structured (**b**) PZT FC/porous polyurethane composites at volume fractions of FC $m = 0.1, 0.2, 0.3$ and 0.4 (reprinted from paper by Khanbareh et al. [30], with permission from Taylor & Francis)

with (1.16). It is probable that the contribution from the piezoelectric coefficient d_{31}^{*} to the dielectric permittivity $\varepsilon_{33}^{*\sigma}$ would be less considerable as a result of the system of aligned FC inclusions poled along the OX_3 axis.

An important addition to the studied volume-fraction dependence of the dielectric permittivity $\varepsilon_{33}^{*\sigma}$ dependence is concerned with results by Krivoruchko [33]. He first interpreted the $\varepsilon_{33}^{*\sigma}$ (m) dependence of non-structured and DEP-structured 0–3 $PbTiO_3$-type FC/epoxy composites [34] in terms of a model of a 0–3 composite with aligned spheroidal FC inclusions; the composite model applied the method for evaluations are described in Sect. 5.1. An agreement between the calculated and experimental volume-fraction dependences is achieved in the following cases:

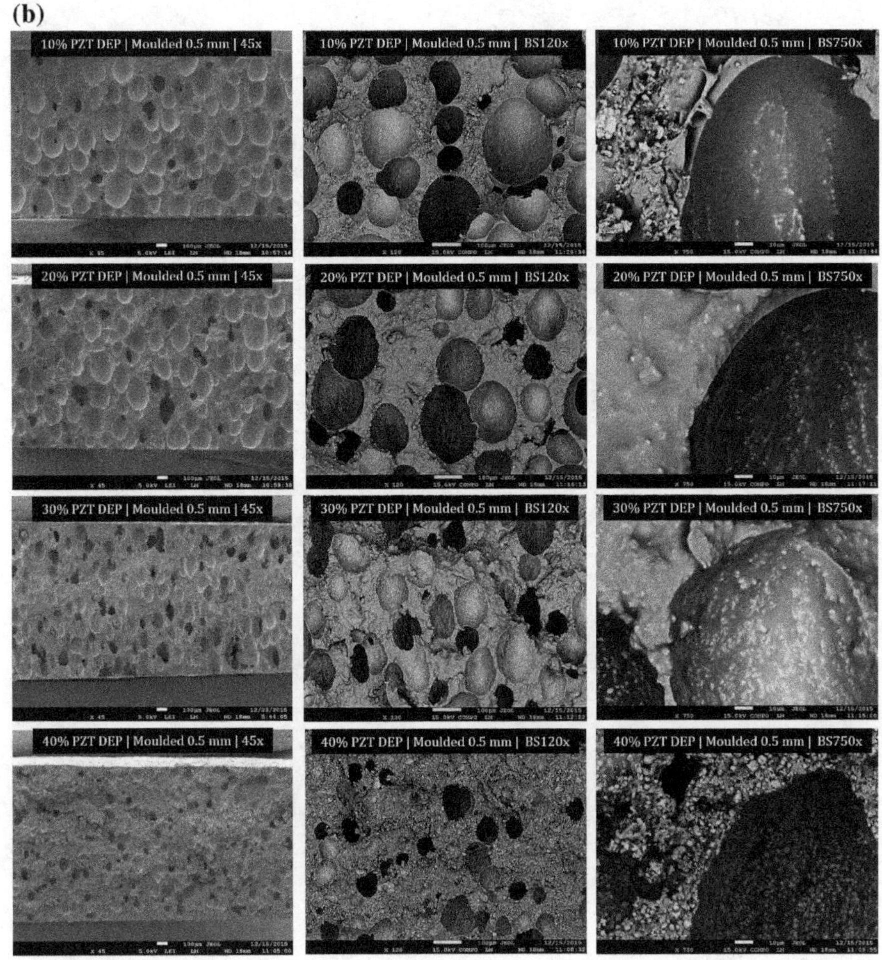

Fig. 3.13 (continued)

(i) in the structured composite, there are poled prolate FC inclusions with an aspect ratio (i.e. ratio of the semi-axes of the spheroidal inclusion) from approximately 0.13 to 0.19 and with a volume fraction of such inclusions in the volume-fraction range of $0 < m \leq 0.4$, and

(ii) in the non-structured composite at $0 < m \leq 0.4$, the aspect ratio of the similar prolate FC inclusions can vary from approximately 0.32 to 0.50 [33].

Figure 3.14b suggests that DEP structuring of the 0–3 composite leads to an increase of the d_{33}^* value by about two–four times in the studied volume-fraction range of $0 < m \leq 0.4$. It should be noted that the configuration of the curves in Fig. 3.14b, c is analogous to the configuration of the curves in Fig. 1.1. We remind the reader that the volume-fraction behaviour of the piezoelectric coefficients d_{33}^*

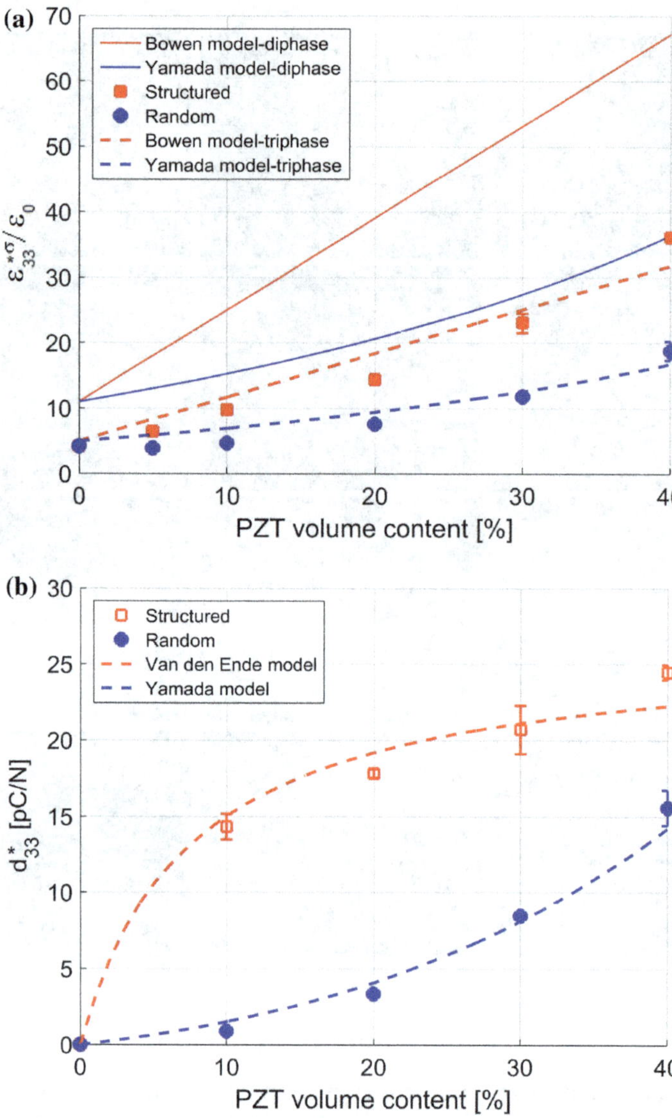

Fig. 3.14 Volume-fraction dependences of the dielectric (**a**) and piezoelectric (**b**–**d**) properties of the DEP-structured and random PZT FC/porous polyurethane composites at room temperature [15]. The piezoelectric coefficient g_{33}^* has been evaluated in accordance with (2.2) by using data from graphs a and b and. Prediction results for the aforementioned volume-fraction dependences obtained within the framework of a few models of composites are compared to experimental data (dots) in the PZT volume-fraction range of $0 \leq m \leq 0.4$

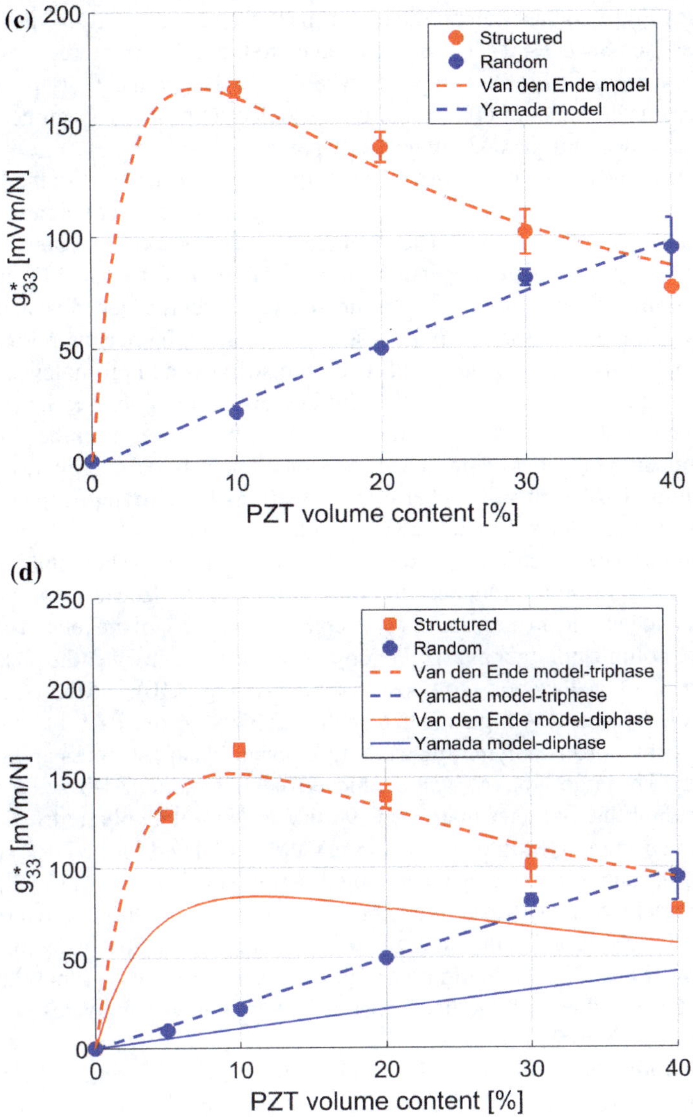

Fig. 3.14 (continued)

and g_{33}^* (see Fig. 1.1) has been predicted for the 0–3 and 1–3 $PbTiO_3$-type FC/epoxy composites (see Sect. 1.4.2). The curves related to the DEP-structured composite in Fig. 3.14b, c are built for the composite wherein the 1–3 connectivity pattern plays a key role. For the random composite, its piezoelectric properties can be modelled by taking into account the 0–3 connectivity pattern.

Changes in the elastic properties of the polyurethane matrix due to the formation of the porous foam structure [30] can influence the piezoelectric coefficient d_{33}^* of the DEP-structured composite [15]. The piezoelectric response of the composite along the poling OX_3 axis strongly depends on the distribution of the poled FC inclusions and their alignment along the OX_3 axis, and, to a lesser degree, depends on the elastic properties of the matrix surrounding these FC inclusions. Increasing porosity in the polymer matrix leads to a decrease of its elastic stiffness and promotes an increase of d_{33}^* in comparison to the composite with the monolithic polymer matrix [7, 35]. It should be noted that the elastic stiffness of the polymer matrix in the studied FC-based composites provides a minimal absorption of mechanical energy by this matrix upon loading. This results in higher effective strains in the composite sample and influences its piezoelectric response [15] to a large extent.

Data on the piezoelectric coefficient g_{33}^* from Fig. 3.14c, d show that the porous polymer matrix strongly influences the longitudinal piezoelectric sensitivity of the composite and leads to an increase of g_{33}^* in comparison to g_{33}^* of the related composite with the monolithic polymer matrix. Hereby we see advantages over the piezoelectric coefficient g_{33}^* of the PT/epoxy composite (see data in Fig. 3.10), and these advantages are accounted for by the higher piezoelectric activity of the PZT FC and by the presence of the porous polymer matrix with a lower elastic stiffness and dielectric permittivity. The largest $\max g_{33}^*$ value achieved in the random PZT-based composites with the monolithic polymer matrices is 94 mV m/N (at the volume fraction of FC $m = 0.4$), and the largest $\max g_{33}^* = 166$ mV m/N is related to the DEP-structured PZT-based porous composite (at the volume fraction of FC $m = 0.1$) [15, 30]. At $m = 0.1$, in the analogous DEP-structured composite with a monolithic polymer matrix, $\max g_{33}^* = 80$ mV m/N, i.e. the longitudinal piezoelectric sensitivity decreases more than by two times [30]. It should be added for further comparison that the $\max g_{33}^*$ values found for other well-studied structured FC/polymer composites [14, 36] are in a range from 70 to 90 mV m/N.

The diagrams shown in Fig. 3.15 summarise results [15, 30] on the dielectric and piezoelectric performance of the studied DEP-structured PZT-based composites at $m = 0.1$, 0.2, 0.3 and 0.4. Hereby the parameters of the related composites with the monolithic (dense) and porous matrices are compared. Data in Fig. 3.15 suggest that the piezoelectric coefficient g_{33}^* of the composite with a porous matrix is larger than g_{33} of the FC component by about six times. This and other results are to be taken into account when selecting novel composites for piezoelectric sensor, energy harvesting and other applications.

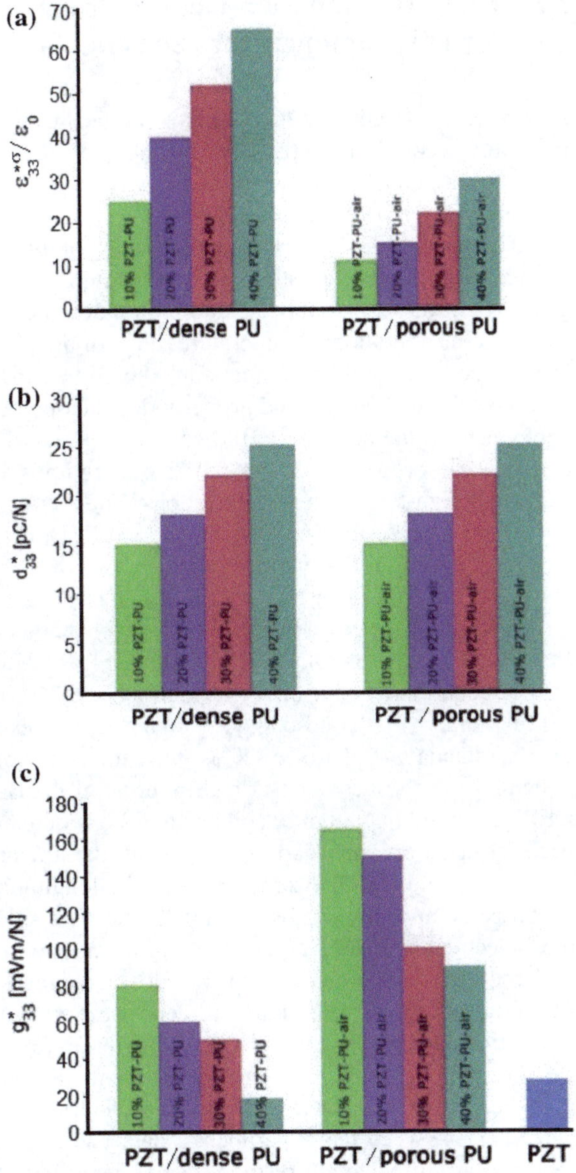

Fig. 3.15 Diagrams that show largest values of the dielectric permittivity $\varepsilon_{33}^{*\sigma}$ (**a**) and piezoelectric coefficients d_{33}^* (**b**) and g_{33}^* (**c**) of the DEP-structured PZT FC/dense polyurethane and PZT FC/porous polyurethane composites at volume fractions of FC $m = 0.1, 0.2, 0.3$ and 0.4. The piezoelectric coefficient g_{33} of the monolithic PZT FC is shown with the rectangle on the right side of the diagram **c** (reprinted from paper by Khanbareh et al. [30], with permission from Taylor & Francis)

3.5 Dielectric and Piezoelectric Properties of Composites for High-Temperature Sensing

As a rule, exploitation of piezo-active composites in high-temperature applications is limited because of the reduced sensing capability of the poled FC component at elevated temperatures, increased electrical conductivity and mechanical attenuation, and due to a temperature dependence of the piezoelectric properties of the FC [37, 38]. The relatively low Curie temperature T_C of many FCs [9–11, 13] restricts their applications in high-temperature ranges, often at $T > 500$ K. In addition, the low glass transition temperature T_g and high thermal expansion of the polymer matrix can limit the operational temperature range of the piezo-active composites [14, 15]. In practice, the operating temperature should be substantially below T_C in order to minimise thermal ageing and property degradation of the FC as a key piezoelectric component in the composite. In Sect. 3.5, we describe the effective dielectric and piezoelectric properties of some PT-based composites [39] that are suitable for a variety of high-temperature applications. The method to manufacture the PT-based composites is described in Sect. 2.6 and work [15, 39].

The effect of DEP structuring on the dielectric permittivity $\varepsilon_{33}^{*\sigma}$ of the PT-based composites is shown in Fig. 3.16a. The PT FC is characterised by a high Curie temperature that is close to $T_C = 765$ K for the single-crystal PT sample [9–11]. PEI is fully imidised ODPA/ODA, a piezo-passive polymer component prepared [15] in accordance with the schematic in Fig. 2.8. The experimental data on $\varepsilon_{33}^{*\sigma}$ for the random PT-based composites can be well explained within the framework of the Yamada's model (see [27]), while the behaviour of $\varepsilon_{33}^{*\sigma}$ in the related DEP-structured composites is described in terms of the Bowen's model (see [32]). The DEP-structured FC/polymer composite is considered [32] as a collection of the FC particles that are aligned and forming chain-like structures along the poling direction.

The influence of DEP structuring on the longitudinal piezoelectric effect in the PT-based composites is shown in Fig. 3.16b. The volume-fraction behaviour of the piezoelectric coefficient d_{33}^* is described in terms of the Yamada's model (see [27]) and van den Ende's model (see [28, 36]). An interfacial interaction between the FC particles and polymer matrix influences the effective dielectric and piezoelectric properties of the piezo-active composite materials [39, 40]. A lack of interfacial adhesion between the FC and polymer components in the composite leads to a decrease in the dielectric permittivity $\varepsilon_{33}^{*\sigma}$ and piezoelectric coefficient d_{33}^* [41, 42]. The good agreement between the experimental data and predictions (see Fig. 3.16) can be regarded as an indication of the effective interfacial adhesion and pore-free interfaces, and this is consistent with SEM images of the microstructures of the studied composite samples [39].

Examples of the temperature dependence of the piezoelectric properties of the PT-based composites are shown in Fig. 3.17. We observe the relatively high thermal stability of the piezoelectric coefficients d_{33}^* and g_{33}^* in both the random and structured composites. To a large extent, this thermal stability is a result of the thermally stable dielectric and piezoelectric properties of the PT in the studied temperature range

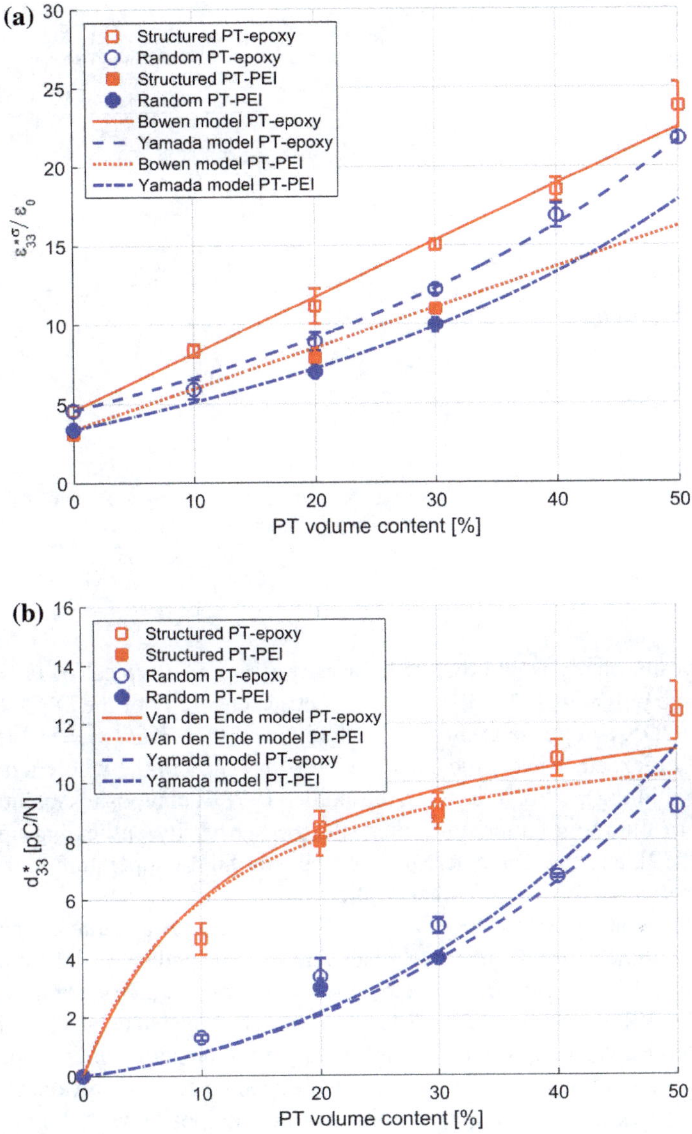

Fig. 3.16 Dielectric (**a** [15]) and piezoelectric (**b** [15] and **c**) properties of DEP-structured and random PT-based composites (figure **c** reprinted from paper by Khanbareh et al. [39], with permission from the Royal Society of Chemistry)

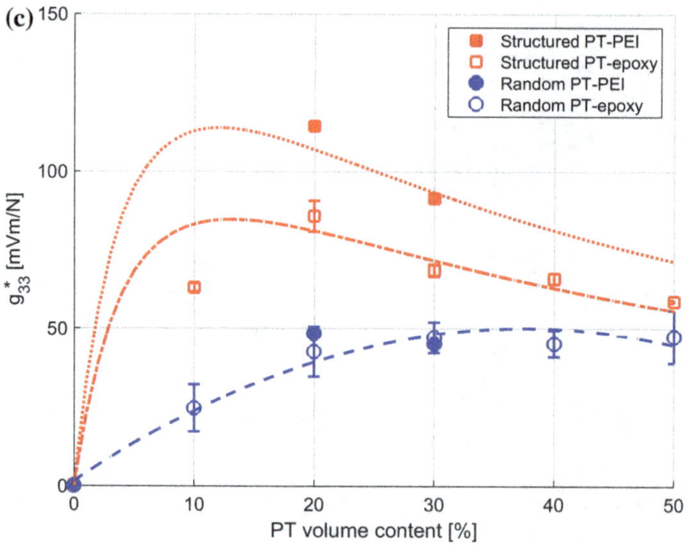

Fig. 3.16 (continued)

[9–11]. The thermally stable dielectric properties are also typical of ODPA/ODA and PEI-type polymers [15, 43, 44]. Piezoelectric sensitivity of the DEP-structured PT/ODPA/ODA composite at the volume fraction of the PT FC $m = 0.2$ is characterised by a piezoelectric coefficient $g_{33}^* > 100$ mV m/N in the whole temperature range shown in Fig. 3.17b. In general, the studied PT/PEI composites maintain more than 85% of the room-temperature piezoelectric sensitivity on heating up to $T = 450$ K [15, 39], and this characteristic is of value at high-temperature piezoelectric sensing, energy harvesting and other applications.

It should be added that in contrast to ODPA/ODA with a stable dielectric permittivity in a wide temperature range, many ferroelectric and piezoelectric polymers (e.g. PVDF and PVDF—co-trifluoroethylene) show a rapid decrease of the dielectric permittivity [45] on heating at $T > 370$ K. This circumstance may be a barrier for the related composites on the way to the high-temperature piezo-technical applications.

Finally, as follows from the g_{33}^* (m) dependence [46] evaluated for a 0–3 $(Pb_{0.9625}La_{0.025})(Ti_{0.99}Mn_{0.01})O_3$ FC/araldite composite with highly prolate spheroidal inclusions oriented along the poling axis OX_3, values of $g_{33}^* \approx 100$ mV m/N at the volume fraction $m = 0.1$–0.3 are approximately equal to those shown for the DEP-structured PT-based composite in Fig. 3.16c. This approximate equality means that

(i) changes in the FC composition concerned with modified PT do not lead to drastic changes in the g_{33}^* (m) dependence related to the composite,
(ii) differences between the elastic properties of polymers in the PT-based composites do not lead to considerable changes in the g_{33}^* value at the volume fraction of FC $m = 0.1$–0.3, and

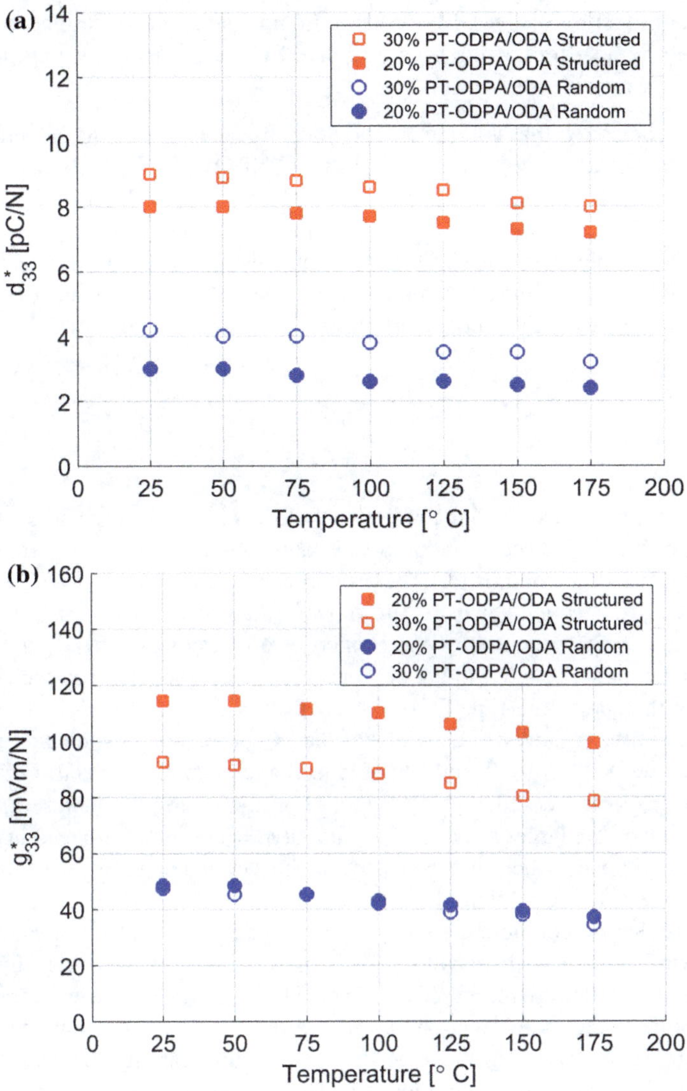

Fig. 3.17 Temperature dependence of piezoelectric coefficients d_{33}^* and g_{33}^* of DEP-structured and random PT-based composites (reprinted from paper by Khanbareh et al. [39], with permission from the Royal Society of Chemistry)

(iii) the architecture of the DEP-structured composite can be in general approximated by a system of the highly prolate FC inclusions [5, 46] aligned in the polymer matrix.

It should be added that volume-fraction dependences of the piezoelectric coefficient d_{33}^{*} and dielectric permittivity $\varepsilon_{33}^{*\sigma}$ of a 0–3 $PbTiO_3$-type FC/polymer composite with two types of FC inclusions (spherical and prelate cylindrical) were discussed in work [47]. Hereby the polymer component of the studied composite was either piezoelectric or pies-passive. As follows from results [47], the prelate FC inclusions aligned in the sample determine to a considerable extent the piezoelectric and dielectric properties of the studied composite at volume fractions of FC $m \geq 0.2$.

3.6 Conclusion

In Chap. 3, we have described the effective dielectric, pyroelectric and piezoelectric properties of the DEP-structured composites and compared these properties to those of the related random composites. The properties of the random and DEP-structured composites are determined by using the known experimental methods. The influence of the composite microgeometry and properties of the components on the effective properties are also discussed in the context of the examples. Hereby the composite is based on the perovskite-type FC – either PT or PZT.

Changes in the hysteresis behaviour of the FC-based composite lead to changes in its properties. As follows from data in Figs. 3.4 and 3.5, the dielectric permittivity $\varepsilon_{33}^{*\sigma}$ and piezoelectric coefficient d_{33}^{*} increase on increasing the volume fraction of FC m. Moreover, the larger $\varepsilon_{33}^{*\sigma}$ and d_{33}^{*} values at $m = \text{const}$ are achieved in the DEP-structured composite wherein the system of the FC chains (or prolate inclusions) aligned on the poling direction plays the key role.

The pyroelectric coefficient $p = p_{3}^{*\sigma}$ and related figure of merit F from (3.2) increase with an increase in the volume fraction m in the composite, see Figs. 3.6 and 3.7. These parameters are improved due to the DEP structuring of the composite and, therefore, due to the more favourable orientation of the FC inclusions with respect to the poling axis OX_3. The range of the temperature stability of the pyroelectric coefficient $p_{3}^{*\sigma}$ of the studied composite is determined.

Our analysis of the piezoelectric sensitivity of the manufactured composites shows that the larger value of maximum of the piezoelectric coefficient g_{33}^{*} describing this sensitivity is achieved in the DEP-structured composite at $m = \text{const}$. The system of the aforementioned FC chains promotes a larger piezoelectric coefficients d_{33}^{*} and g_{33}^{*} in the DEP-structured composite. The considerable increase of the piezoelectric coefficient g_{33}^{*} is associated with forming a porous structure in the polymer matrix of the composite. Examples of the influence of the porous (foam) polymer matrix on the piezoelectric coefficient g_{33}^{*} are considered for the PZT FC/porous polyurethane composite. The system of well-distributed spherical pores (polyurethane foam) leads to the decrease of the dielectric permittivity of the polymer matrix and, therefore, to the decrease of the dielectric permittivity $\varepsilon_{33}^{*\sigma}$ of the composite. The highly favourable combination of DEP structuring of PZT particles and the porous polymer matrix in the three-component composite leads to its improved piezoelectric sensitivity: in this case, the piezoelectric coefficient $g_{33}^{*} \approx 100$ mV m/N and max$g_{33}^{*} = 166$ mV m/N are achieved. This result means that the piezoelectric coefficient g_{33}^{*} concerned with the longitudinal response of the studied composite can be approximately four—seven times larger than the piezoelectric coefficient $g_{33}^{(1)}$ of an analogous PZT FC system. In addition, a few micromechanical models were used to interpret the experimental volume-fraction dependences of the effective properties of the manufactured piezo-active composites.

Diagrams are put forward (Fig. 3.15) wherein the largest values of the dielectric permittivity $\varepsilon_{33}^{*\sigma}$ and piezoelectric coefficients d_{33}^{*} and g_{33}^{*} of the DEP-structured PZT FC/dense polyurethane and PZT FC/porous polyurethane composites are shown for a volume fraction of FC $m = \text{const}$. These diagrams illustrate the advantages of the studied three-component PZT-based composite with a porous polymer matrix over the related two-component composite with the monolithic matrix.

High-temperature PT-based composites exhibit the thermal stability of both the dielectric and piezoelectric properties (see Fig. 3.17). This stability is observed in the presence of FC with the high Curie temperature T_C and due to the considerable thermal stability of the dielectric properties of the polymer matrix.

It is believed that the experimental data shown and analysed in Chap. 3 will be effective at the selection of novel composites for piezo-technical, pyroelectric, energy harvesting and other important applications.

References

1. L.P. Khoroshun, B.P. Maslov, P.V. Leshchenko, *Prediction of Effective Properties of Piezo-Active Composite Materials* (Naukova Dumka, Kiev, 1989) (in Russian)
2. V.M. Levin, M.I. Rakovskaja, W.S. Kreher, The effective thermoelectroelastic properties of microinhomogeneous materials. Int. J. Solids Struct. **36**, 2683–2705 (1999)
3. Yu.V. Sokolkin, A.A. Pan'kov, *Electroelasticity of Piezo-Composites with Irregular Structures* (Fizmatlit, Moscow, 2003) (in Russian)
4. R. Kar-Gupta, T.A. Venkatesh, Electromechanical response of piezoelectric composites: effects of geometric connectivity and grain size. Acta Mater. **56**, 3810–3823 (2008)
5. V.Yu. Topolov, C.R. Bowen, *Electromechanical Properties in Composites Based on Ferroelectrics* (Springer, London, 2009)
6. V.Yu. Topolov, P. Bisegna, C.R. Bowen, *Piezo-active Composites. Orientation Effects and Anisotropy Factors* (Springer, Heidelberg, New York, Dordrecht, London, 2014)
7. V.Yu. Topolov, C.R. Bowen, P. Bisegna, *Piezo-Active Composites. Microgeometry – Sensitivity Relations* (Springer International Publishing, Cham, 2018)
8. R.E. Newnham, D.P. Skinner, L.E. Cross, Connectivity and piezoelectric-pyroelectric composites. Mater. Res. Bull. **13**, 525–536 (1978)
9. B. Jaffe, W.R. Cook, H. Jaffe, *Piezoelectric Ceramics* (Academic Press, London New York, 1971)
10. Y. Xu, *Ferroelectric Materials and Their Applications* (North-Holland, Amsterdam, London, New York, Toronto, 1991)

11. A.V. Gorish, V.P. Dudkevich, M.F. Kupriyanov, A.E. Panich, A.V. Turik, *Piezoelectric Device-Making. Vol. 1: Physics of Ferroelectric Ceramics* (Radiotekhnika, Moscow, 1999) (in Russian)
12. P. Greil, Mikrostruktur keramischer Werkstoffe, ed. by Keramik, H. Schaumburg (B. G. Teubner, Stuttgart, 1994), pp. 29–104
13. K. Ruschmeyer, G. Helke, J. Koch, K. Lubitz, T. Möckl, A. Petersen, M. Riedel, A. Schönecker, *Piezokeramik: Grundlagen, Werkstoffe, Applikationen* (Expert-Verlag, Renningen-Malmsheim, 1995)
14. H. Khanbareh, S. van der Zwaag, W. Groen, Effect of dielectrophoretic structuring on piezoelectric and pyroelectric properties of PT-epoxy composites. Smart Mater. Struct. **23**, 105030 (2014)
15. H. Khanbareh, Expanding the functionality of piezo-particulate composites. Proefschrift ter verkrijging van der grad van doctor aan de Technische Universiteit Delft (Delft, 2016)
16. I.S. Zheludev, *Physics of Crystalline Dielectrics. Vol. 2: Electrical Properties* (Plenum, New York, 1971)
17. S.B. Lang, Pyroelectricity: from ancient curiosity to modern imaging tool. Phys. Today **58**, 31–36 (2005)
18. R.W. Whatmore, R. Watton, Pyroelectric materials and devices, in *Infrared Detectors and Emitters: Materials and Devices*, ed. by P. Capper, C.T. Elliott (Springer, New York, 2001), pp. 99–147
19. C.R. Bowen, J. Taylor, E. Le Boulbar, D. Zabek, V.Yu. Topolov, A modified figure of merit for pyroelectric energy harvesting. Mater. Lett. **138**, 243–246 (2015)
20. C. Wong, F. Shin, Effect of electrical conductivity on poling and the dielectric, pyroelectric and piezoelectric properties of ferroelectric 0–3 composites. J. Mater. Sci. **41**, 229–249 (2006)
21. W.K. Sakamoto, D. Kanda, D. Das-Gupta, Dielectric and pyroelectric properties of a composite of ferroelectric ceramic and polyurethane. Mater. Res. Innovations **5**, 257–260 (2002)
22. W. Sakamoto, P. Marin-Franch, D. Das-Gupta, Characterization and application of PZT/PU and graphite doped PZT/PU composite. Sens. Actuators, A **100**, 165–174 (2002)
23. P. Marin-Franch, T. Martin, D. Tunnicliffe, D. Das-Gupta, PTCa/PEKK piezo-composites for acoustic emission detection. Sens. Actuators, A **99**, 236–243 (2002)
24. P. Marin-Franch, D. Tunnicliffe, D. Das-Gupta, Pyroelectric properties of the PTCa/PEKK composite transducers, in *Eighth International Conference on Dielectric Materials, Measurements and Applications* (IEE Conf. Publ. No. 473). 17–21 September 2000, Edinburgh, UK. (Institution of Electrical Engineers, 2000), pp. 386–391
25. M. Abdullah, D. Das-Gupta, Electroactive properties of polymer-ceramic composites. Ferroelectrics **87**, 213–228 (1988)
26. S. Bravina, N. Morozovsky, J. Kułek, B. Hilczer, Pyroelectric thermowave probing and polarization reversal in TGS/PEO composites. Mol. Cryst. Liq. Cryst. **497**, 109/[441]–120/[452] (2008)
27. T. Yamada, T. Ueda, T. Kitayama, Piezoelectricity of a high-content lead zirconate titanate/polymer composite. J. Appl. Phys. **53**, 4328–4332 (1982)
28. D. van den Ende, B. Bory, W. Groen, S. van der Zwaag, Improving the d_{33} and g_{33} properties of 0–3 piezoelectric composites by dielectrophoresis. J. Appl. Phys. **107**, 024107 (2010)
29. H. Savakus, K. Klicker, R. Newnham, PZT-epoxy piezoelectric transducers: a simplified fabrication procedure. Mater. Res. Bull. **16**, 677–680 (1981)
30. H. Khanbareh, K. de Boom, S. van der Zwaag, W.A. Groen, Highly sensitive piezo particulate-polymer foam composites for robotic skin application. Ferroelectrics **515**, 25–33 (2017)
31. V.Yu. Topolov, A.V. Turik, Porous piezoelectric composites with extremely high reception parameters. Tech. Phys. **46**, 1093–1100 (2001)
32. C.P. Bowen, R.E. Newnham, C.A. Randall, Dielectric properties of dielectrophoretically assembled particulate-polymer composites. J. Mater. Res. **13**, 205–210 (1998)
33. A.V. Krivoruchko, Effects of combination of physical properties and orientation effects in ferroelectric composites. Abstract of thesis, Cand. Sci. (Phys. & Math.) (Voronezh, 2009) (in Russian)

34. S.A. Wilson, G.M. Maistros, R.W. Whatmore, Structure modification of 0–3 piezoelectric ceramic/polymer composites through dielectrophoresis. J. Phys. D Appl. Phys. **38**, 175–182 (2005)
35. S. Van Kempen, Optimisation of piezoelectric composite materials design through improved materials selection and property prediction methods, Master's thesis (Delft University of Technology, Delft, 2012)
36. D.A. van den Ende, Structured piezoelectric composites, materials and applications, Ph. D. thesis (Delft University of Technology, Delft, 2012)
37. F. Yu, Q. Lu, S. Zhang, H. Wang, X. Cheng, X. Zhao, High-performance, high-temperature piezoelectric BiB_3O_6 crystals. J. Mater. Chem. C **3**, 329–338 (2015)
38. D. Damjanovic, Ferroelectric, dielectric and piezoelectric properties of ferroelectric thin films and ceramics. Rep. Prog. Phys. **61**, 1267–1324 (1998)
39. H. Khanbareh, M. Hegde, J.C. Bijleveld, S. van der Zwaag, P. Groen, Functionally graded ferroelectric polyetherimide composites for high temperature sensing. J. Mater. Chem. C **5**, 9389–9397 (2017)
40. Y. Shen, Y. Lin, C.-W. Nan, Interfacial effect on dielectric properties of polymer nanocomposites filled with core/shell-structured particles. Adv. Func. Mater. **17**, 2405–2410 (2007)
41. M. Carvalho Araújo, C.M. Costa, S. Lanceros-Méndez, Evaluation of dielectric models for ceramic/polymer composites: effect of filler size and concentration. J. Non-Cryst. Solids **387**, 6–15 (2014)
42. J. Yao, C. Xiong, L. Dong, C. Chen, Y. Lei, L. Chen, R. Li, Q. Zhu, X. Liu, Enhancement of dielectric constant and piezoelectric coefficient of ceramic–polymer composites by interface chelation. J. Mater. Chem. **19**, 2817–2821 (2009)
43. Q. Chi, J. Sun, C. Zhang, G. Liu, J. Lin, Y. Wang, X. Wang, Q. Lei, Enhanced dielectric performance of amorphous calcium copper titanate/polyimide hybrid film. J. Mater. Chem. C **2**, 172–177 (2014)
44. X. Fang, X. Liu, Z.-K. Cui, J. Qian, J. Pan, X. Li, Q. Zhuang, Preparation and properties of thermostable well-functionalized graphene oxide/polyimide composite films with high dielectric constant, low dielectric loss and high strength via in situ polymerization. J. Mater. Chem. A **3**, 10005–10012 (2015)
45. C. Baur, Y. Zhou, J. Sipes, S. Priya, W. Voit, Organic, flexible, polymer composites for high-temperature piezoelectric applications. Energy Harvest. Syst. Mater., Mech., Circuits Storage **1**, 167–177 (2014)
46. S.V. Glushanin, V.Yu. Topolov, A.V. Krivoruchko, Features of piezoelectric properties of 0–3 $PbTiO_3$-type ceramic/polymer composites. Mater. Chem. Phys. **97**, 357–364 (2006)
47. S.V. Glushanin, V.Yu. Topolov, Features of the electromechanical properties of 0–3 composites of the $PbTiO_3$-based ferroelectric ceramics–polymer type. Tech. Phys. Lett. **31**, 346–348 (2005)

Chapter 4
Modelling of the Composite Structure Formation During Dielectrophoresis

Abstract Analytical and computational results concerned with modelling of the piezo-particulate composite structure formation during dielectrophoresis are analysed. A numerical model has been put forward to simulate the ceramic-particle alignment and structure formation in a viscous medium during dielectrophoresis, and the system 'ferroelectric ceramic particles + medium' is assumed to be semi-dilute. The main parameters to be considered during this modelling are the order parameter and total chain perfection degree. These parameters are inseparably linked with some microgeometric characteristics of the system of the ceramic particles in the continuous viscous medium. The performance of the aforementioned model is demonstrated for a variety of systems based on ferroelectric ceramic particles in a viscous polymer medium.

In Chaps. 2 and 3, we have focused our attention on methods suitable to manufacture modern piezo-particulate composites and on their effective properties that are of value in specific applications. As shown in work [1–4], an intermediate state of the composite between 0–3 (FC inclusions in a large polymer matrix, particulate state) and 1–3 (FC rods in a large polymer matrix, fibrous state) can be achieved by the DEP treatment on a semi-dilute solution of the FC particles (or inclusions) in a viscous matrix. Such a viscous state is typical of the most polymer systems prior to curing [1–3, 5]. Under ideal processing conditions, the process of forming the long and well-separated particulate threads is observed, and these threads can span the entire electrode-to-electrode spacing, approaching the topology of the perfect 1–3 FC composites [6, 7]. An alignment of the particles in the thread-like structure at DEP leads to a marked increase in the piezoelectric properties of the composite, while the high mechanical flexibility due to its polymer matrix is maintained [2, 5]. The quality of the FC particle alignment in the thread-like composite structure is the key parameter in controlling its effective [8–11]. Enhancing the alignment quality by means of decreasing the inter-particle distance leads to improving the dielectric, piezoelectric and pyroelectric properties of the particulate composite [2]. An effect of processing parameters on the formation of the threads in the composite has been experimentally demonstrated for the high-aspect-ratio FC particles in a thermoset polymer matrix [12]. A quality of the composite structure formation has been deter-

H. Khanbareh et al., *Piezo-Particulate Composites*, Springer Series in Materials Science 283, https://doi.org/10.1007/978-3-030-19204-4_4

mined after the consolidation of the polymer matrix and additional processing [3]. In this context, it is important to analyse specifics of the formation of the 0–3/1–3 composite structures under an external electric field and its indirect influence on the piezoelectric properties of the composites manufactured by means of DEP.

In this chapter, we discuss analytical and computational results concerned with the modelling of the composite structure formation during DEP [13]. As earlier, we consider examples of the FC/polymer composites manufactured for the piezoelectric transducer, sensor, acoustic and other applications.

4.1 Dielectrophoretic Force

The DEP force is defined [13–15] as an action experienced by a poled dielectric particle in an external electric field $\boldsymbol{E}$. In the ideal case, when this particle can be modelled by means of an infinitesimal dipole moment $\boldsymbol{p}$, an absolute value of the force F that acts on the particle is

$$F = p\nabla E, \tag{4.1}$$

and a torque that acts on the same particle is

$$\boldsymbol{M} = [\boldsymbol{p}\,\boldsymbol{E}]. \tag{4.2}$$

Equations (4.1) and (4.2) are based on the assumption that the electric field $\boldsymbol{E}$ is not influenced by the dipole $\boldsymbol{p}$ [14, 15]. However, at the structure formation during DEP, the aforementioned assumptions of the infinitesimal dipole moment $\boldsymbol{p}$ and on lack of the influence caused by $\boldsymbol{p}$ do not hold. The FC particles are to be considered as finite 'domains' with an own polarisation $\boldsymbol{P}$, and this polarisation would contribute into the electric field $\boldsymbol{E}$ in the surrounding medium. In addition, this contribution will induce a force that acts on the neighbouring FC particles. The described situation can be treated analytically, for instance, by means of so-called higher-order multipoles, see work [14, 16–18]. Alternatively and especially to carry out a simulation of the composite structure formation with several particles involved in this consideration, an emphasis is placed rather on a qualitative behaviour of the system than on accuracy of results and applied numerical methods. Among these methods, the finite element method is of interest because this method provides a meaningful alternative in many respects.

Our study is limited to a two-dimensional rectangular domain that can be treated as an idealisation of a thin three-dimensional quadrangular domain Ω^l filled with a viscous liquid [3, 13]. This liquid is characterised by the dielectric permittivity $\varepsilon^{(l)}$ and viscosity η, confined between two plates separated by a distance d, and containing a single layer of particles $\{\Omega_i^p\}_{i=1}^n$ with the dielectric permittivity $\varepsilon^{(p)}$ and mass density γ restricted to move on the mid-plane. The thickness of the particles is l.

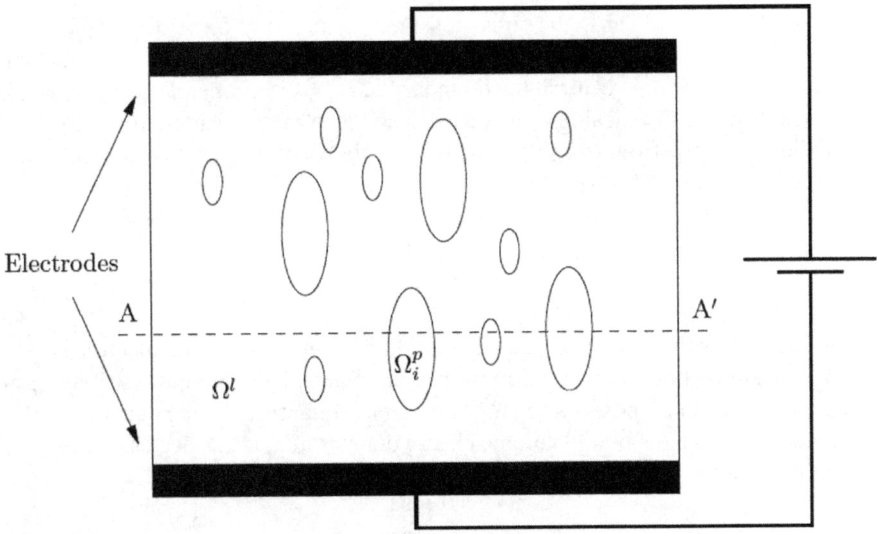

Fig. 4.1 Schematic representation of the model of DEP structuring (reprinted from paper by Gutiérrez et al. [13], with permission from Elsevier)

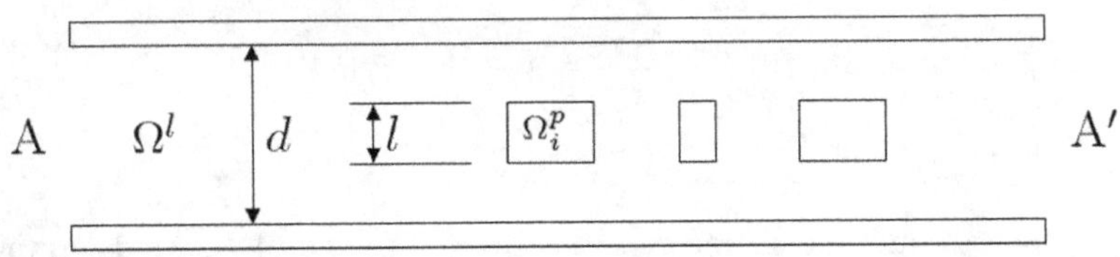

Fig. 4.2 Cross section of the model (see Fig. 4.1) along the AA' line (reprinted from paper by Gutiérrez et al. [13], with permission from Elsevier)

This renders a computationally treatable model of an experimental set-up [5] that is traditionally used to observe the composite structure formation at DEP manufacturing. A schematic of the above-described model is given in Fig. 4.1. For a better understanding of the arrangement, the cross section of the present model is shown in Fig. 4.2.

Hereafter, the FC particles Ω_i^p are modelled as linear elastic solids in plane-stress conditions [13]. The motion of each individual particle is obtained at solving the following Cauchy equation:

$$\nabla \cdot \sigma + b = \ddot{\gamma} u \tag{4.3}$$

in Ω_i^p, and the relation

$$\sigma n = t \tag{4.4}$$

holds on $\partial\Omega_i^p$. In (4.3), σ is the stress tensor, b is the body force density, γ is the mass density and u is the displacement field. In (4.4), t is the boundary traction and $\boldsymbol{n}$ is the outward normal vector at the boundary $\partial\Omega_i^p$. For a proper description of the particle motion, geometrical nonlinearity is also taken into consideration [13].

The body force b from (4.3) consists of the DEP contribution b_d and equivalent viscous drag term $b_{v.}$ The boundary traction t from (4.4) is of the DEP nature only.

The Laplace equation from electrostatics [19]

$$\nabla^2 V = 0 \tag{4.5}$$

is now solved for the electric potential V in the region between the electrodes (see Fig. 4.1). Hereby, boundary conditions are implemented to obtain a distribution of the electric field $\boldsymbol{E}$ and polarisation $\boldsymbol{P}$ from derived quantities. The polarisation $\boldsymbol{P}$ is associated [13] with the bound volume charge density ρ_V and surface charge density Σ_S in the particles according to the relations

$$\rho_V = -\nabla \boldsymbol{P} \tag{4.6}$$

in Ω_i^p and

$$\sum_S = \boldsymbol{P}\,\boldsymbol{n} \tag{4.7}$$

on $\partial\Omega_i^p$.

Taking into account the charge densities from (4.6) and (4.7), we determine the DEP forces that act on the FC particle, from the relation [13]

$$b_d = \rho_V E t = \sum_S E. \tag{4.8}$$

4.2 Viscous Drag

The system of the FC particles $\left\{\Omega_i^p\right\}_{i=1}^n$ in the fluid medium experiences the viscous drag upon the motion. To describe this drag, we use a simplified model [3, 13]. It is assumed that

(i) fluid interacts with the FC particle along the top and bottom flat surfaces, and
(ii) rate of a shear is linear between the particle and the upper and lower confining plates (Fig. 4.3). If the Newtonian type of fluid is considered within the framework of classical mechanics, then a distributed shearing force acting on the surface is

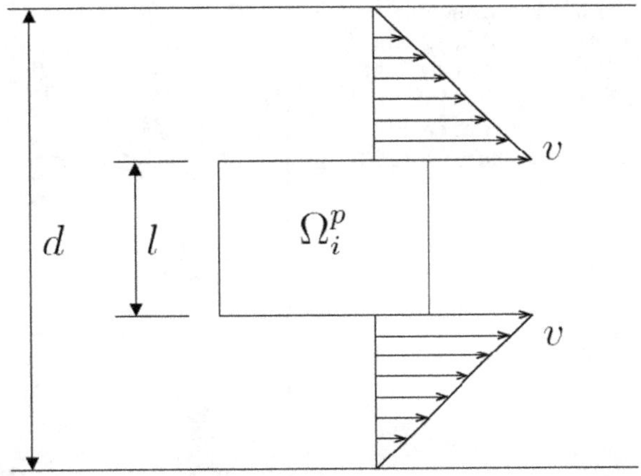

Fig. 4.3 Schematic representation of the assumed velocity field for computation of the viscous drag force

$$\tau = \eta v / [(d - l)/2]. \tag{4.9}$$

Then, the equivalent body force is given by

$$b_v = 2\tau / l, \tag{4.10}$$

and τ is from (4.9). The factor two is introduced in (4.10) to describe the contribution from both the sides shown in Fig. 4.3.

4.3 Modelling and Parameters

The model described in Sect. 4.1 is then simulated in the multiphysics finite element package COMSOL [20] making use of the electromagnetism and the solid mechanics modules [3, 13]. A single finite element domain is generated for the domain $\cup_{i=1}^{n} \Omega_i^p \cup \Omega^l$ respecting the internal boundaries between the FC particles and fluid, see Figs. 4.1 and 4.4. The electrostatic problem is solved in the entire domain to obtain the DEP forces, see details in work [3, 13]. After updating the position of all the FC particles, the mesh in the liquid domain is adapted. For this purpose, a moving mesh technique is adopted in combination with automatic remeshing [21]. At forming the composite structure, the FC particles will eventually collide and become attached. This is modelled by means of contact constraints to prevent a domain overlapping [13].

A quality of the composite structure evolution is studied by means of a set of objective geometric parameters that describe this structure (Fig. 4.4). These parameters are also related to the resulting performance of the studied FC/polymer composite in the poled state [13]. One of the objective parameters which are commonly used to

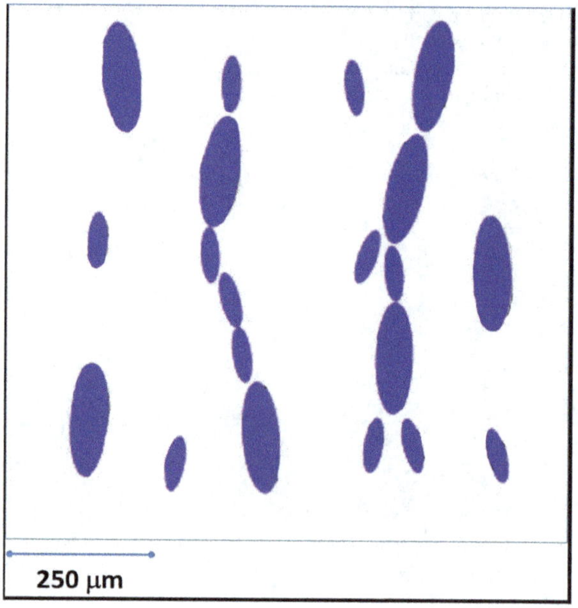

Fig. 4.4 Schematic arrangement of the system of the FC particles in the fluid medium (reprinted from paper by Gutiérrez et al. [13], with permission from Elsevier)

describe an orientation order of an ensemble of the prolate FC particles is the order parameter [22]

$$\overline{P_2} = \overline{\frac{3}{2}\cos^2\beta - \frac{1}{2}} \tag{4.11}$$

In (4.11), β is the angle between the dominating dimension of the particle and the electric field $\boldsymbol{E}$ vector. The order parameter $\overline{P_2}$ from (4.11) can vary from $\overline{P_2} = 0$ for a random arrangement of the FC particles to $\overline{P_2} = 1$ for a system of the fully aligned particles with respect to the $\boldsymbol{E}$ direction. Specifics of the calculation and use of $\overline{P_2}$ can be found in work [3, 13]. The second important parameter that influences the piezoelectric performance of the DEP-structured composite is the inter-particle distance d, i.e. the distance between the closest FC particles within the formed chain. In an optimal situation, all the prolate FC particles form the chains with no gaps in between, and this leads to $d = 0$. It is not feasible to give a formal and objective definition of the parameter d, as it depends on the actual composite structure that has been formed at DEP structuring. Due to this feature, the involved pairs of the FC particles, for which d is computed, need to be identified in the first line. These particles are obviously the same for which contact constraints have been defined during the simulation [13]. The inter-particle distance d is calculated via the Euclidean distance transform of the binary image. For each pixel in the binary image, the distance transform assigns a number that is the distance between that pixel and the nearest nonzero pixel of the binary image. Hereafter, a greyscale image is produced wherein every pixel within a feature is assigned a value that is its straight line distance from the nearest background pixel. This is analogous to a dilatation of the particles with constant speed until the growing border meets the border of the neighbouring particles. Subsequently, on each merging point between every two particles, two distance values are identified with respect to both particles within a chain. The sum of these two values is the final inter-particle distance [3].

A quality of the chains formed during DEP is described in terms of a chain perfection degree. This parameter is introduced for a topological comparison of the formed composite structure to its ideal analogue, i.e. to a system of the perfect chains spanning from one electrode to the other and leading to the 1–3 connectivity pattern. To evaluate the perfection degree of the chain, we should detect the chain as an element of the composite structure. Using the image processing toolbox, the snapshots are transformed into binary images, followed by calculation of the sum of pixels in every column. Finally, the sum value is plotted versus the column position, and peaks in the resulting spectrum correspond to the chain position. The inter-columnar distance between the peaks shows the lateral distance between the chains formed in the composite [3, 13]. The total chain perfection degree PD_c is expressed in terms of a global perfection degree PD_g and a local perfection degree PD_l:

$$PD_c = PD_g\ PD_l = [P_{2c}(L-d)/L_{ic}]\left[\overline{P_2}(1-(\Sigma d/a))\right] \tag{4.12}$$

In (4.12), P_{2c} is the order parameter of the chain, L is the actual chain length, L_{ic} is the ideal chain length, d is the inter-particle distance and $\overline{P_2}$ is the average order parameter of the individual particles within the studied chain.

4.4 Interaction Between Ceramic Particles and Their Influence on the Composite Structure

Poled FC particles located in the external electric field $\boldsymbol{E}$ experience a torque $\boldsymbol{M}$ from (4.2) in a case of the infinitesimal dipole idealisation. The orientation behaviour is studied for the FC particles with various aspect ratios, by evaluating the $\overline{P_2}(t)$ dependence. The time $t = 0$ corresponds to the initial configuration of the polarisation axis of the FC particle at 45° with respect to the $\boldsymbol{E}$ direction. Results on $\overline{P_2}(t)$ [3] show that the aspect ratio of the FC particles has a minor influence on the alignment behaviour of the system of such particles in the electric field $\boldsymbol{E}$.

At the two-particle interaction, two versions of the evolution of the composite structure can be considered [3, 13]. In the first version, it is assumed that two prolate FC particles are symmetrically aligned with the electric field $\boldsymbol{E}$ at a mutual distance being equal to the particle length (as measured along the $\boldsymbol{E}$ vector). Upon poling, these particles will repel each other in the horizontal direction, see the evolution in Fig. 4.5. This example shows that the centrosymmetrically aligned particles barely interact with each other once they are separated by a finite distance.

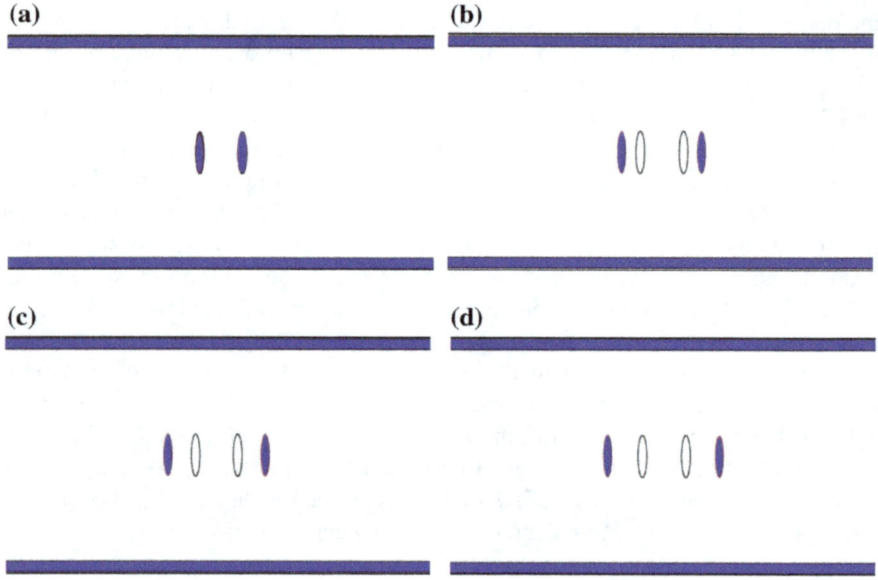

Fig. 4.5 Positional evolution for two centrosymmetrically aligned particles at $t = 0$ s (**a**), $t = 0.5$ s (**b**), $t = 1.0$ s (**c**) and $t = 1.5$ s (**d**). Empty and filled ellipsoids show the initial and actual positions of the particles, respectively. The external electric field $\boldsymbol{E}$ is acting on the vertical direction (reprinted from paper by Gutiérrez et al. [13], with permission from Elsevier)

The similar simulation is carried out for the same particles, but now placed initially in an off-centred position (Fig. 4.6). In the electric field $\boldsymbol{E}$, the particles will not only 'feel' the imposed external homogenous field but also the local field induced around each particle. As a consequence, these two particles first rotate towards each other, but away from the equilibrium orientation angle. The rotation of these particles is accompanied by a lateral and axial displacement that reduces the inter-particle distance d. Both motions lead to the formation of the connected chains at an off-axis angle with respect to the imposed electric field $\boldsymbol{E}$.

To minimise the total energy of the system 'particles – chains – polymer medium', the chain rotates to align with the external field $\boldsymbol{E}$ (Fig. 4.6c, d). The present example shows that the rotation of the prolate particles precedes their lateral or axial displacement in accordance with experimental results [5, 23]. In general, the chain formation during DEP does not only depend on the polymer properties and applied electric field $\boldsymbol{E}$, but also on the initial configuration of the FC particles with respect to each other.

The next example of the formation of the chain is concerned with interaction and evolution of nine prolate FC particles in the electric field $\boldsymbol{E}$ [3, 13]. Features of such an evolution are described in terms of the inter-particle distance d and orientation parameter $\overline{P_2}$ that changes with the time t. In the initial configuration, the particles are placed on a 3 × 3 regular array. After the application of the electric field $\boldsymbol{E}$ and

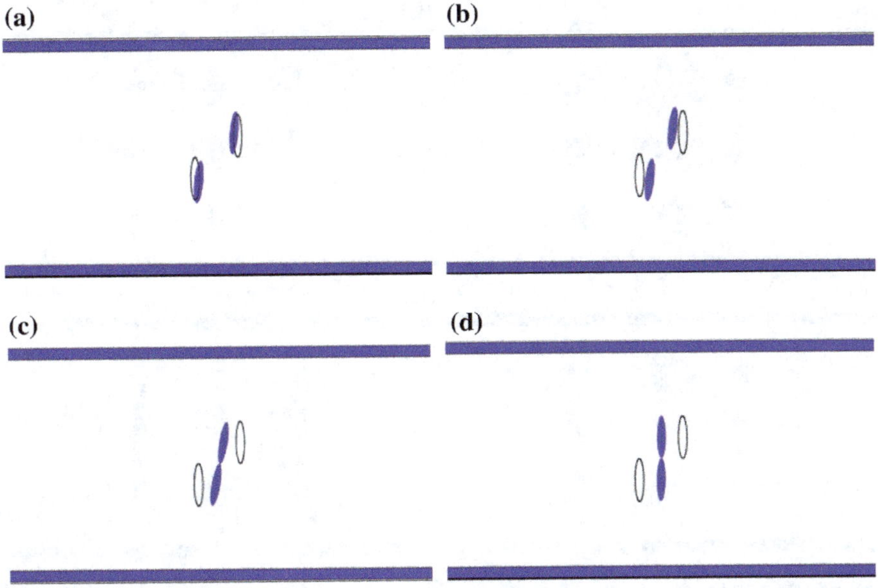

Fig. 4.6 Positional evolution for two parallel particles positioned off-centre at $t = 0.2$ s (**a**), $t = 0.4$ s (**b**), $t = 0.5$ s (**c**) and $t = 0.6$ s (**d**). Empty and filled ellipsoids show the initial and actual positions, respectively. The external electric field $\boldsymbol{E}$ is acting on the vertical direction (reprinted from paper by Gutiérrez et al. [13], with permission from Elsevier)

once all particles have aligned on the $\boldsymbol{E}$ direction (at $t = 0.1$ s, Fig. 4.7a), they start interacting until two chains of three particles and one chain of two particles have formed (see Fig. 4.7b–d).

In order to evaluate the chain quality, a chain identification process is applied to the four configurations shown in Fig. 4.7, and the results are shown in Fig. 4.8. In Fig. 4.8, we see the sum of the pixel values in every column (see the ordinate axis), and the curves are plotted versus the column position (see the abscissa axis, arbitrary units). Each peak shown in Fig. 4.8 corresponds to a chain of particles, and the inter-columnar distance between the peaks means the lateral distance between the chains formed during DEP. The time (t) scale of the FC-component alignment is comparable with that of a single particle [13]; however, it remains smaller than that related to the chain formation, as follows from the $\overline{P_2}(t)$ dependence shown in Fig. 4.9. According to (4.11), the $\overline{P_2}(t)$ dependence is caused by changes in the angle β between the dominating dimension of the particle and the external electric field $\boldsymbol{E}$. The present simulation of the composite structure formation during DEP is consistent with experimental results of the time-dependent evolution of the average orientation angle and average chain length, see work [23]. We note that, in accordance with Fig. 4.9, $\overline{P_2}(t)$ does not increase continuously. At $t > 0.1$ s, the particles have attained an initial alignment (see Fig. 4.7b–d); therefore, $\beta(t)$ becomes almost constant, and we observe the saturation of $\overline{P_2}(t)$ in Fig. 4.9. At the same time, there is a minor

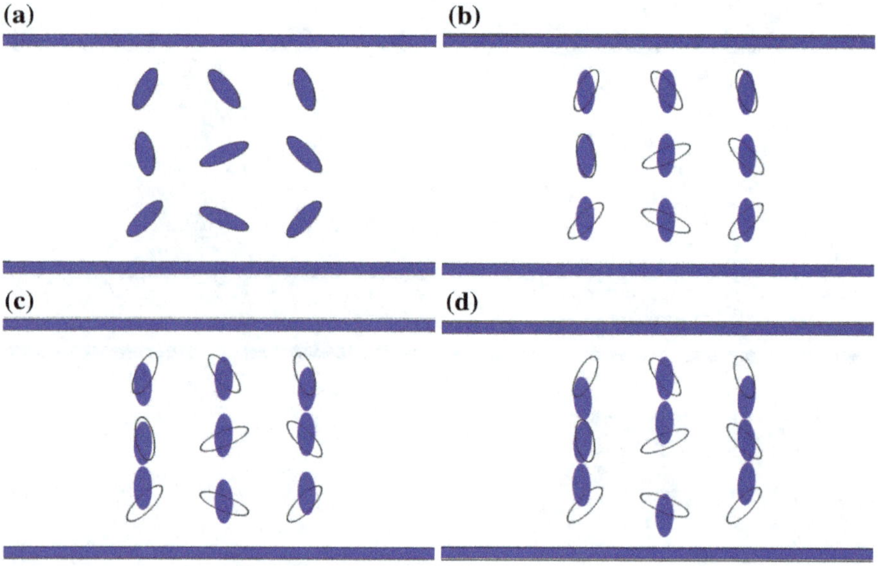

Fig. 4.7 Evolution in terms of nine identical particles that were initially placed on a simple square lattice: $t = 0$ s (**a**), $t = 0.1$ s (**b**), $t = 0.4$ s (**c**) and $t = 0.7$ s (**d**). Empty and filled ellipsoids show the initial and actual positions of these particles, respectively. The external electric field $\boldsymbol{E}$ is acting on the vertical direction (reprinted from paper by Gutiérrez et al. [13], with permission from Elsevier)

fluctuation in the $\overline{P_2}$ value as the particles experience minor temporary displacements [3] from the well-determined sites.

The inter-particle distance d_c as a function of time t is graphically represented in Fig. 4.10. The character of the decreasing $d_c(t)$ dependence suggests that it is sufficient to consider that all particles are aligned at the initial configuration, and the further evolution of the system of particles does not play an important role. Moreover, the final inter-particle distance d_c approaches zero (see Fig. 4.10), i.e. the particles tend to be well connected within the chain oriented along the $\boldsymbol{E}$ vector. This behaviour agrees with experimental results [2, 3] and shows that, irrespective of the volume fraction of FC, the inter-particle distance in the DEP-structured composite samples remains very small in comparison with linear sizes of in individual FC grain.

A system of the prolate particles with two different aspect ratios is shown in Fig. 4.11. Due to the specifics of the distribution of the internal electric field in this system, the more prolate particles act as a guide which predetermines the positions of the less prolate particles. In Fig. 4.11c, d, we observe that the parallel particles repel each other to follow the $\boldsymbol{E}$ vector, and this behaviour resembles the positional evolution for the two parallel particles shown in Fig. 4.6c, d.

An evolution in the system of the randomly distributed large and small particles in the electric field $\boldsymbol{E}$ during DEP is shown in Fig. 4.12. The larger particle creates a more intensive local field, and this promotes the formation of the long chain along

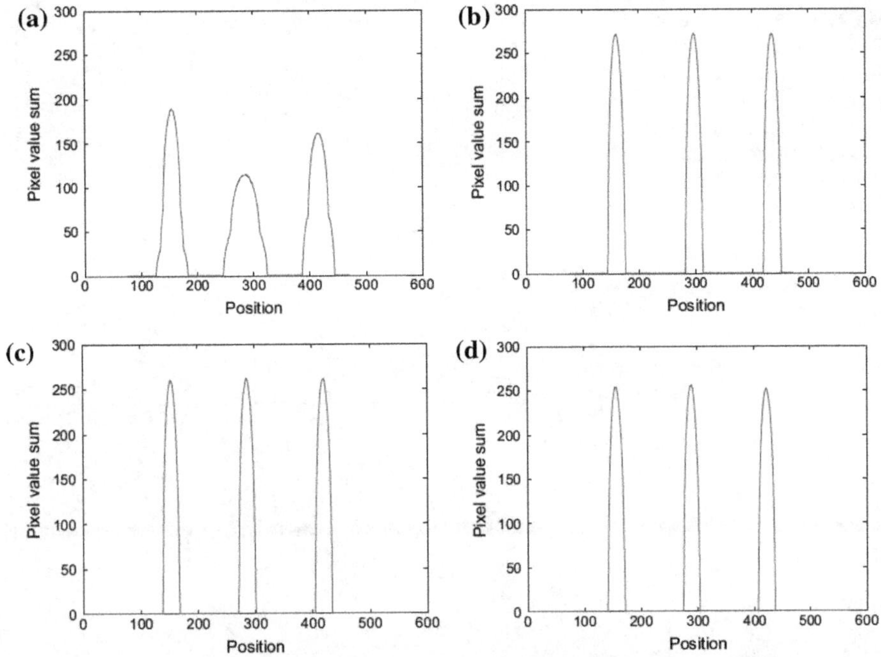

Fig. 4.8 Chain identification process for the system of nine identical particles shown in Fig. 4.7. Graphs **a–c** and **d** are related to $t = 0$ s, 0.1 s, 0.4 s and 0.7 s, respectively (reprinted from paper by Gutiérrez et al. [13], with permission from Elsevier)

Fig. 4.9 Evolution of the $\overline{P_2}$ parameter from (4.11) for the system of the nine identical particles shown in Fig. 4.7 (reprinted from paper by Gutiérrez et al. [13], with permission from Elsevier)

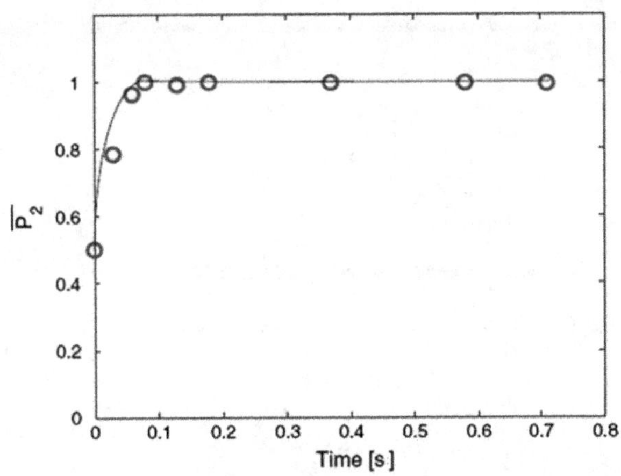

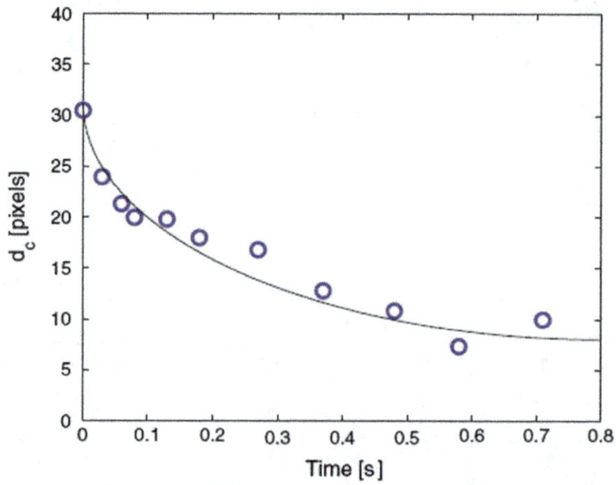

Fig. 4.10 Dependence of the average inter-particle distance d_c on the time for the system of the nine identical particles shown in Fig. 4.7 (reprinted from paper by Gutiérrez et al. [13], with permission from Elsevier)

(a) (b) (c) (d)

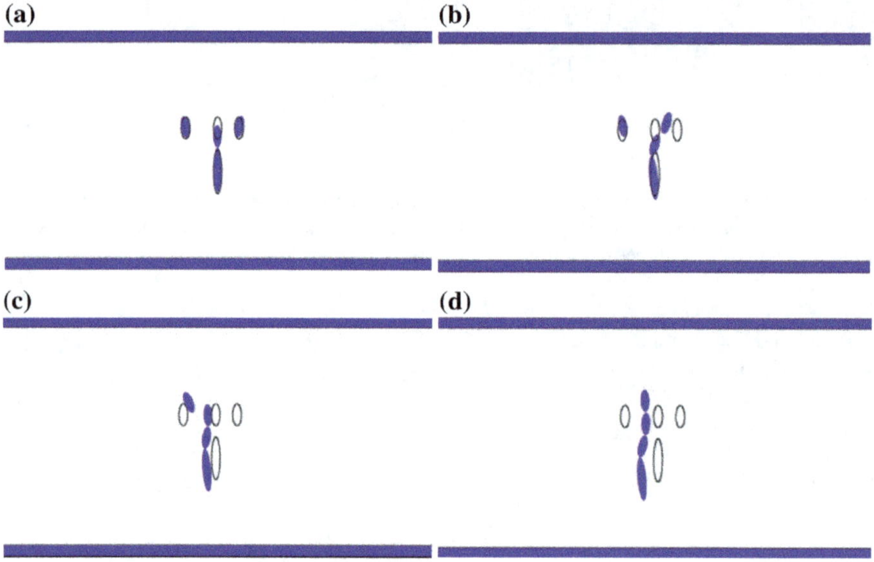

Fig. 4.11 Evolution in the system of particles with two different aspect ratios at $t = 0.05$ s (**a**), $t = 0.15$ s (**b**), $t = 0.35$ s (**c**) and $t = 0.50$ s (**d**). Empty and filled ellipsoids show the initial and actual positions of the particles, respectively. The external electric field $\boldsymbol{E}$ is acting on the vertical direction (reprinted from paper by Gutiérrez et al. [13], with permission from Elsevier)

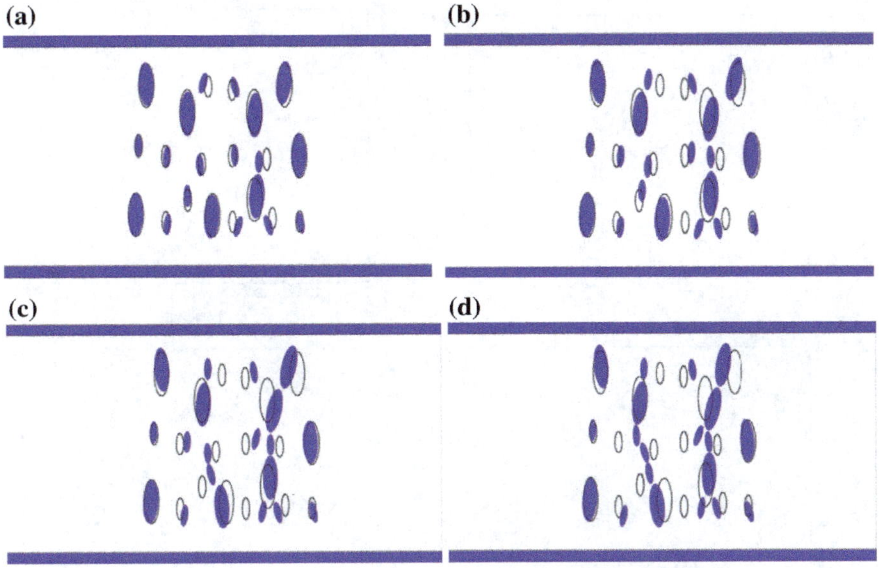

Fig. 4.12 Evolution in the system of particles with different sizes at $t = 0.035$ s (**a**), $t = 0.070$ s (**b**), $t = 0.100$ s (**c**) and $t = 0.125$ s (**d**). Empty and filled ellipsoids show the initial and actual positions of the particles, respectively. The external electric field ***E*** is acting on the vertical direction (reprinted from paper by Gutiérrez et al. [13], with permission from Elsevier)

the ***E*** vector. After the application of an electric field ***E***, the small FC particles tend to fill in the gaps between the large particles to form a continuous chain.

The chain identification process in the present system is described with graphs in Fig. 4.13. Hereby, one can draw a restricted analogy between the graphs in Figs. 4.8 and 4.13. The main reason for the difference between the graphs shown in Figs. 4.8 and 4.13 is a result of the distribution of the FC particles on their sizes, and this is to be taken into account at further modelling at DEP structuring. As follows from Figs. 4.12 and 4.13, four chains are identified at times $t \geq 0.1$ s. For $t < 0.1$ s, the individual chains cannot be properly identified (see Fig. 4.13a, b), and the chain perfection degree cannot be calculated [13].

4.5 Conclusion

In this chapter, we have analysed some features of DEP structuring and modelling of the formation of the composite structure in the 0–3 and 1–3 FC/polymer composites. A numerical model has been put forward [2, 3] to simulate the particle alignment and structure formation in a viscous medium during DEP, and the system 'FC particles + medium' is assumed to be semi-dilute. The main parameters to be considered during modelling are the order parameter $\overline{P_2}$ from (4.11) and the total chain perfection degree

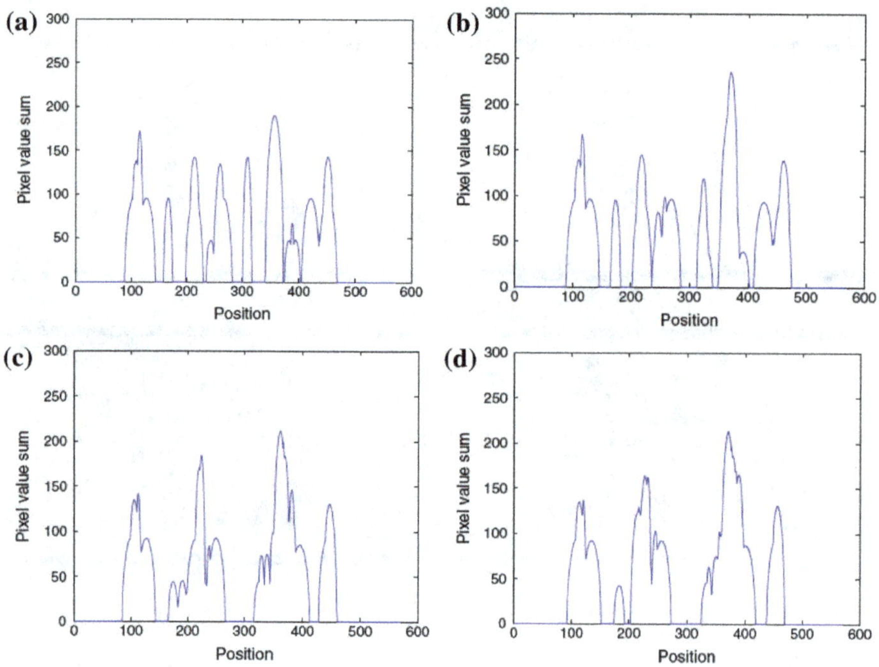

Fig. 4.13 Chain identification process for the system of particles with different sizes, see Fig. 4.12. Graphs **a**–**c** and **d** are related to $t = 0.035$ s, 0.070 s, 0.100 s and 0.125 s (**d**), respectively (reprinted from paper by Gutiérrez et al. [13], with permission from Elsevier)

PD_c from (4.12). These parameters are inseparably linked with some microgeometric characteristics of the system of the FC particles in the continuous viscous medium.

The performance of the present model of the formation of the composite structure is demonstrated for a variety of systems based on FC particles in a viscous medium. The general behaviour trend of the studied heterogeneous system after application of the electric field $\boldsymbol{E}$ is summarised as follows.

(i) the reorientation of the FC particles occurs with respect to the imposed electric field $\boldsymbol{E}$, and due to the orientation effect stimulating the field $\boldsymbol{E}$, a thread or chain formation proceeds in the system. The present model gives a quantitative handle to analyse a submodel performance in experimental situations. The effect of the FC particle aspect ratio is minor when considering the alignment behaviour of a single-particle system. The initial configuration of a particle pair has been observed to be a key parameter in controlling the particle interaction. The repulsion is observed when a pair of particles is symmetrically aligned with the electric field, resulting in bare interaction after separation by a finite distance between these particles. However, the chain formation is realised if the neighbouring particles are initially located at off-centred positions.

(ii) the effect of the FC particle size and size distribution is studied for a multiparticle system. The particle-size distribution is regarded as a key parameter at the chain formation during DEP. Larger particles act as a guide for smaller particles, which consequently dictate the kinetics and final degree of the orientation of the whole system of the particles with respect to the imposed electric field $\boldsymbol{E}$.

The results described and discussed in Chap. 4 are to be used at the prediction of features of DEP structuring in systems of FC particles with specific chemical compositions, shapes and sizes. Some results on DEP structuring and microgeometric features can be taken into consideration at the further interpretation of the effective physical properties in the manufactured FC/polymer composites.

References

1. S.A. Wilson, G.M. Maistros, R.W. Whatmore, Structure modification of 0–3 piezoelectric ceramic/polymer composites through dielectrophoresis. J. Phys. D Appl. Phys. **38**, 175–182 (2005)
2. H. Khanbareh, S. van der Zwaag, W. Groen, Effect of dielectrophoretic structuring on piezoelectric and pyroelectric properties of PT-epoxy composites. Smart Mater. Struct. **23**, 105030 (2014)
3. H. Khanbareh, Expanding the functionality of piezo-particulate composites. Proefschrift ter verkrijging van der grad van doctor aan de Technische Universiteit Delft. (Delft, 2016)
4. H. Khanbareh, S. van der Zwaag, W.A. Groen, In-situ poling and structurization of piezoelectric particulate composites. J. Intell. Mater. Syst. Struct. **28**, 2467–2472 (2017)
5. D. van den Ende, B. Bory, W. Groen, S. van der Zwaag, Improving the d_{33} and g_{33} properties of 0–3 piezoelectric composites by dielectrophoresis. J. Appl. Phys. **107**, 024107 (2010)
6. H.L.W. Chan, J. Unsworth, Simple model for piezoelectric ceramic/polymer 1–3 composites used in ultrasonic transducer applications. IEEE Trans. Ultrason. Ferroelectr. Freq. Control **36**, 434–441 (1989)
7. V.Yu. Topolov, C.R. Bowen, *Electromechanical Properties in Composites Based on Ferroelectrics* (Springer, London, 2009)
8. H. Tang, M.H. Malakooti, H.A. Sodano, Relationship between orientation factor of lead zirconate titanate nanowires and dielectric permittivity of nanocomposites. Appl. Phys. Lett. **103**, 222901 (2013)
9. S.K. Patil, M.Y. Koledintseva, R.W. Schwartz, W. Huebner, Prediction of effective permittivity of diphasic dielectrics using an equivalent capacitance model. J. Appl. Phys. **104**, 074108 (2008)
10. C. Andrews, Y. Lin, H.A. Sodano, The effect of particle aspect ratio on the electroelastic properties of piezoelectric nanocomposites. Smart Mater. Struct. **19**, 025018 (2010)
11. H. Tang, Z. Zhou, H.A. Sodano, Relationship between $BaTiO_3$ nanowire aspect ratio and the dielectric permittivity of nanocomposites. ACS Appl. Mater. Interfaces. **6**, 5450–5455 (2014)
12. D.A. van den Ende, Structured piezoelectric composites, materials and applications. Ph.D. thesis. Delft University of Technology, Delft, 2012
13. M. Gutiérrez, H. Khanbareh, S. van der Zwaag, Computational structure formation during dielectrophoresis in particulate composites. Comput. Mater. Sci. **112**, 139–146 (2016)
14. T.B. Jones, Multipole corrections to dielectrophoretic force. IEEE Trans. Ind. Appl. IA-21, 930–934 (1985)
15. T.B. Jones, Dielectrophoretic force in axisymmetric fields. J. Electrostat. **18**, 55–62 (1986)
16. T.B. Jones, *Electromechanics of Particles* (Cambridge University Press, Cambridge New York Melbourne Madrid Cape Town Singapore São Paulo, 2005)

17. N. Green, T.B. Jones, Numerical determination of the effective moments of nonspherical particles. J. Phys. D Appl. Phys. **40**, 78–85 (2007)
18. B. Techaumnat, B. Eua-arporn, T. Takuma, Electric field and dielectrophoretic force on a dielectric particle chain in a parallel-plate electrode system. J. Phys. D Appl. Phys. **37**, 3337–3346 (2004)
19. L.D. Landau, E.M. Lifshitz, *Electrodynamics of Continuous Media* (Pergamon, Oxford, 1984)
20. COMSOL, Inc., COMSOL Multiphysics™User's Guide (version 3.4) (2007). http://www.comsol.com
21. E. Stein, R. de Borst, T.J.R. Hughes, *Encyclopedia of Computational Mechanics*, 2nd edn. (Wiley, Chichester, 2018)
22. N.D. Spencer, J.H. Moore (eds.), *Encyclopedia of Chemical Physics and Physical Chemistry* (Institute of Physics Publishing, Bristol, 2001)
23. D.A. van den Ende, H.J. van de Wiel, W.A. Groen, S. van der Zwaag, Direct strain energy harvesting in automobile tires using piezoelectric PZT-polymer composites. Smart Mater. Struct. **21**, 015011 (2012)

Chapter 5
Prediction of Effective Properties of Composites Based on Ferroelectric Ceramics

Abstract The important step at the study of the piezo-particulate composites is concerned with the prediction of their effective properties. Hereby a number of models have been put forward at modelling and interpretation of the properties. Examples of the effective piezoelectric properties and related parameters of 0–3, 1–3 and 2–2 ferroelectric ceramic/polymer composites, and 0–3-type composites with two kinds of the ceramic inclusions are discussed to demonstrate the influence of microgeometric characteristics, components and other factors on the electromechanical coupling and piezoelectric performance of the composites. The effective pyroelectric and dielectric properties of the ceramic-based composites are considered to a lesser degree. The microgeometry of the piezo-active composite plays a key role in determining its piezoelectric sensitivity, and changes in the microgeometry can lead to appreciable changes in the piezoelectric sensitivity and related parameters. Main examples of the piezoelectric sensitivity are concerned with composites that are based on either the PZT-type or $PbTiO_3$-type ceramics. Calculated parameters are compared to the known experimental data.

The main connectivity patterns considered in the present monograph are 0–3 and 1–3. As is known from work [1] and Table 1.1, a microgeometry of the 0–3 composite corresponds to a system of isolated inclusions in a large continuous matrix, and in a 1–3 composite these inclusions are infinitely long along one of the co-ordinate axes, i.e. they form a system of aligned rods. In the 0–3 and 1–3 composites, as a rule, the piezoelectric inclusions exhibit higher properties in comparison to the surrounding matrix [1–7]. The well-known piezo-active 0–3 and 1–3 composites [2–5] contain FC and polymer components whereby the poling direction of the FC component coincides with the poling direction of the composite as a whole, and the polymer component is often piezo-passive. In recent decades, the piezo-active FC/polymer composites with 0–3, 1–3 and 2–2 connectivities have been studied by means of various experimental [2–6, 8–10] and theoretical [3, 11–25] methods. In Chapter 5, we analyse examples of the piezoelectric properties and related parameters of the 0–3, 1–3 and 2–2 FC-based composites that can be of interest across a variety of piezo-technical applications.

H. Khanbareh et al., *Piezo-Particulate Composites*, Springer Series in Materials Science 283, https://doi.org/10.1007/978-3-030-19204-4_5

5.1 0–3 Connectivity Patterns and Properties of Composites

5.1.1 Connectivity and Effective Properties

The 0–3 FC/polymer composite can be represented as a three-dimensionally connected polymer matrix reinforced by a system of isolated FC inclusions. Normally, the FC inclusions play a key role in forming the effective electromechanical properties of the 0–3 composite [4, 14–18, 23]; and their shape, distribution, orientation, poling degree, etc. [8, 9, 19] are to be taken into account to describe the performance of the composite. In Fig. 5.1, we illustrate a fragment of a 0–3 composite with a spheroidal inclusion.

The 0–3 FC/polymer composite can be made in a variety of forms [26–28] with a range of mechanical flexibility, and this is to be taken into account, for example, in sensor applications [5]. Modern DEP-based methods [6, 29, 30] and CAD-based methods concerned with rapid prototyping [5, 28] are suitable to manufacture 0–3 FC/polymer composites systems based on aligned FC inclusions and with predictable electromechanical properties. The most common FC components of the 0–3 composites studied in the last three decades are perovskite-type compositions based on Pb(Zr, Ti)O_3 [5, 13, 14, 17, 23] or Pb(Ca, Ti)O_3 [27, 31]. The full sets of electromechanical constants of the perovskite-type FCs in the poled state are given in Table 1.2.

The effective electromechanical properties of the 0–3 FC/polymer composites were predicted and analysed in many papers; see, for instance, work [13–19, 23, 24]. The most popular models for predicting the effective properties often consider a system of regularly shaped inclusions, and their shape can be a sphere, spheroid or parallelepiped. An approach developed in work [19, 32] is useful to analyse the effective properties of 0–3 composites with an irregular arrangement of the FC inclusions in the polymer matrix.

An averaging procedure that allows for an electromechanical interaction [14, 23, 24, 33] between the piezo-active inclusions in the 0–3 FC/polymer composite is often based on the effective field method. This method is the well-known Mori–Tanaka method [34] generalised for piezo-active composites [23, 24] of the matrix type. In the effective field method, an individual inclusion surrounded by a continuous matrix, which is under an external electroelastic field, is subjected to the action of an unknown average electroelastic field that is effective in point of fact. The effective field is considered as a result of the external field applied to the matrix and, therefore, to the 0–3 composite material wherein the electroelastic interaction between the inclusions is taken into account. Using the equivalent inclusion approach, Dunn and Taya [35] modelled the coupled electroelastic behaviour of 0–3 composites based on FCs. This modelling can be performed for a composite with aligned ellipsoidal (or spheroidal) piezoelectric inclusions [35] at their volume fraction from the range of $0 < m \leq 0.5$. In the volume-fraction range of $0.5 \leq m < 1$, results of calculations by means of the effective field method [14] can only be regarded as approximate since

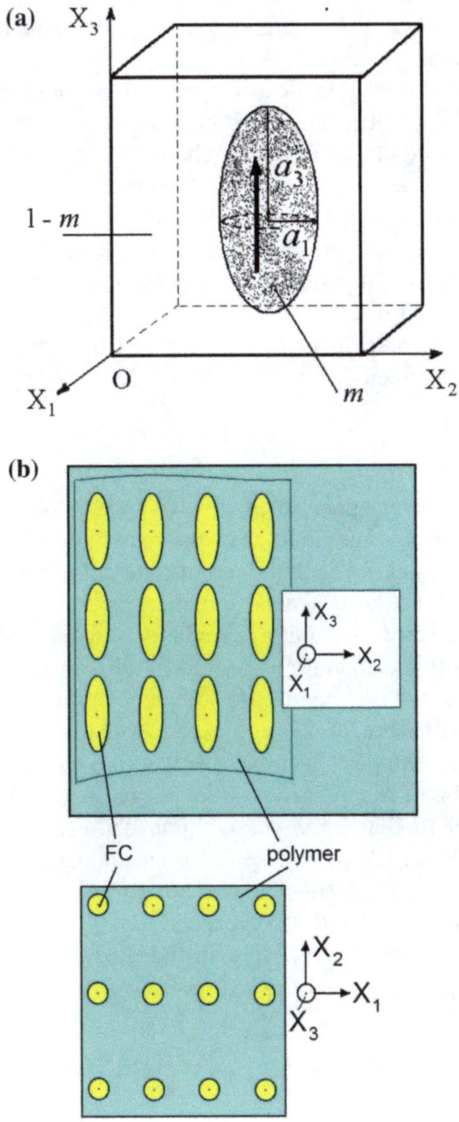

Fig. 5.1 Schematic of the 0–3 composite with spheroidal inclusions (**a**) and arrangement of spheroidal FC inclusions in a polymer matrix (**b**). $(X_1X_2X_3)$ is the rectangular co-ordinate system. In figure a, the polarisation vector of the inclusion is denoted by an arrow, a_i are semi-axes of the spheroid, m is the inclusion volume fraction, and $1 - m$ is the matrix volume fraction (figure **a** reprinted from monograph by Topolov and Bowen [23], with permission from Springer, and figure **b** reprinted from paper by Topolov et al. [25], with permission from Elsevier)

an obvious error due to near-touching or touching of the piezoelectric inclusions increases as their volume fraction $m \to 1$.

Different schemes of averaging are also applied to predict the effective properties of piezo-active 0–3 composites, and in this context we mention the dilute approach, effective medium method and differential scheme [17, 23, 36]. Of specific interest is the FEM [24, 37] applied to piezo-active 0–3 composites with spheroidal inclusions. Results obtained within the framework of different methods are compared and discussed in work [17, 23, 24]. Below we show an application of the effective field method [18, 23–25] to find the effective properties and related parameters of some 0–3 FC/polymer composites.

The shape of the spheroidal inclusion of the 0–3 FC/polymer composite (Fig. 5.1) in the rectangular co-ordinate system $(X_1X_2X_3)$ is described by the equation

$$(x_1/a_1)^2 + (x_2/a_1)^2 + (x_3/a_3)^2 = 1. \tag{5.1}$$

In (5.1), the semi-axes of the spheroid are $a_1 = a_2$ and a_3. We define the aspect ratio of the spheroid as $\rho = a_1/a_3$. We assume that the spheroidal inclusions are uniformly aligned in a continuous matrix, so that the remanent polarisation vector of each FC inclusion is oriented along the OX_3 axis, and this axis is also the poling axis for the composite sample as a whole. Centres of symmetry of the inclusions occupy sites of a simple tetragonal lattice with unit-cell vectors parallel to the OX_j axes. To achieve an improved poling of the 0–3 composite sample, electrical conductivities $\gamma^{(n)}$ of the FC ($n = 1$) and polymer ($n = 2$) components obey the inequality $\gamma^{(2)} \geq \gamma^{(1)}$. It is assumed that surface charges appearing owing to the ferroelectric polarisation of the inclusions are to be fully screened by free charge carriers that flow to interfaces (Fig. 5.1a) between the FC and polymer components. In the averaging procedure, we use (1.4) and (1.5) for the piezoelectric (or electroelastic) medium. We denote the effective properties of the composite with an asterisk '*', and the properties of the FC and polymer with '(1)' and '(2)', respectively.

Following results of work [18, 23], we write the effective electromechanical properties of the 0–3 composite in a general matrix form as follows:

$$\|C^*\| = \begin{pmatrix} \|c^{*E}\| & \|e^*\|^t \\ \|e^*\| & -\|\varepsilon^{*\xi}\| \end{pmatrix}. \tag{5.2}$$

In (5.2), the superscript t denotes the transposition, $\|c^{*E}\|$ is the 6×6 matrix of elastic moduli at electric field $E = \text{const}$, $\|e^*\|$ is the 3×6 matrix of piezoelectric coefficients, and $\|\varepsilon^{*\xi}\|$ is the 3×3 matrix of dielectric permittivities at mechanical strain $\xi = \text{const}$. The aforementioned electromechanical constants are also involved in (1.4) and (1.5).

Taking into account the electromechanical interaction between the aligned piezoelectric inclusions, we find the effective properties in the 0–3 FC/polymer composite by means of the effective field method [18, 23, 33]. The 9×9 matrix of the effective electromechanical properties of the composite $\|C^*\|$ is given by

$$||C*|| = \left||C^{(2)}\right|| + m\left(\left||C^{(1)}\right|| - ||C^{(2)}||\right)\left[||I|| + (1-m)\,||S||\left||C^{(2)}\right||^{-1}\left(\left||C^{(1)}\right|| - \left||C^{(2)}\right||\right)\right]^{-1}. \quad (5.3)$$

In (5.3), the matrices of the electromechanical constants $||C^{(1)}||$ (FC component) and $||C^{(2)}||$ (polymer component) have a form similar to that shown in (5.2), m is the volume fraction of the FC component, $||I||$ is the identity matrix, and $||S||$ is the matrix that contains the Eshelby tensor components. Elements of $||S||$ depend [33, 38] on the elements of $||C^{(2)}||$ (i.e. the electromechanical properties of the medium surrounding the FC inclusion) and the aspect ratio ρ of the FC inclusion. The polymer component can be either piezo-passive (isotropic) or piezo-active (transversely isotropic). The 0–3 FC/polymer composite poled along the OX_3 axis (see Fig. 5.1) is described by ∞mm symmetry, and the effective properties of this composite in (5.2) and (5.3) depend on m and ρ. It should be noted that the effective properties in (5.3) are regarded as homogenised properties in a longwave approximation [22–25], i.e. on assumption that a wavelength of an external acoustic field is much longer than the larger axis of the inclusion.

Based on the effective electromechanical properties found from (5.3) and taking into account relations (1.12)–(1.19) for the piezoelectric medium, we find the remaining piezoelectric coefficients d^*_{ij}, g^*_{ij} and h^*_{ij} from the following matrices:

$$||d^*|| = ||e^*||\left||c^{*E}\right||^{-1},\ ||g^*|| = ||d^*||\,||\varepsilon^{*\sigma}||^{-1} \text{ and } ||h^*|| = ||e^*||\,||\varepsilon^{*\xi}||^{-1}. \quad (5.4)$$

In (5.4), $||\varepsilon^{*\sigma}||$ is the 3 × 3 matrix of dielectric permittivities at mechanical stress σ = const, and $||\varepsilon^{*\xi}||$ the similar matrix of dielectric permittivities at mechanical strain ξ = const. It should be added that electromechanical constants of the 0–3 Pb(Zr, Ti)O_3-type FC/polymer composite which are calculated by means of the effective field method [see (5.3)] are in full agreement with the similar constants of the related 0–3 composite studied in work [14].

5.1.2 0–3 Composites Based on $PbTiO_3$-Type Ceramics

The first example of the piezoelectric performance is concerned with the 0–3 composites based on $PbTiO_3$-type FCs. It is of importance to note that the piezoelectric coefficients $e^{(1)}_{3j}$ of these FCs [18, 23, 32] obey the conditions

$$e^{(1)}_{31} > 0,\ e^{(1)}_{33} > 0 \text{ and } e^{(1)}_{15} > 0, \quad (5.5)$$

and the large anisotropy of their piezoelectric coefficients $d^{(1)}_{3j}$ and ECFs $k^{(1)}_{3j}$ is also observed (see, for instance, the validity of the condition $d^{(1)}_{33}/\left|d^{(1)}_{31}\right| \gg 1$ for modified $PbTiO_3$ and related FCs in Table 1.2).

In Fig. 5.2, we show examples of the calculated volume-fraction dependences of the effective piezoelectric coefficients of a 0–3 composite based on a modified $PbTiO_3$ FC. The important feature of the 0–3 composite based on the $PbTiO_3$-type FC with piezoelectric coefficients $e_{3j}^{(1)}$ that obey (5.5) in the presence of $\min d_{31}^{*}(m, \rho)|_{\rho=\text{const}}$ (see curve 3 in Fig. 5.2a), $\min g_{31}^{*}(m, \rho)|_{\rho=\text{const}}$ and $\max g_{33}^{*}(m, \rho)|_{\rho=\text{const}}$ (see curves 5 and 6 in Fig. 5.2a, b) where ρ is taken from a range of aspect ratios $0 < \rho < \rho^{*}$ that are related to the heavily prolate FC inclusions [23]. The m_f values ($f = 2$ and 3), that obey conditions $g_{31}^{*}(m_2, \rho) = g_{31,m}^{*}$ and $g_{33}^{*}(m_3, \rho) = g_{33,m}^{*}$, are almost equal, and an inequality $|m_2 - m_3| \leq 0.01$ holds. The calculated $e_{33}^{*}(m, \rho)$, $d_{33}^{*}(m, \rho)$ and $g_{3j}^{*}(m, \rho)$ dependences (see curves 2, 4–6 in Fig. 5.2) are similar to a variety of 0–3 FC/polymer composites from work [13, 14, 23]. The $g_{33,m}^{*}$ values related to the studied composite at $\rho \ll 1$ are highly dependent on the elastic moduli of the polymer component [23].

The second example is concerned with the piezoelectric performance of the 0–3 $(Pb_{1-x}Ca_x)TiO_3$ FC/polyurethane composites [18, 23]. The poled $(Pb_{1-x}Ca_x)TiO_3$ FC in the molar concentration range of $0.10 \leq x \leq 0.30$ (Table 5.1) exhibits non-monotonic dependences of both the piezoelectric anisotropy factors $e_{33}^{(1)}/e_{31}^{(1)}$ and $d_{33}^{(1)}/d_{31}^{(1)}$ and an almost constant piezoelectric coefficient $g_{33}^{(1)}$ (Table 5.1). As a consequence, the local maxima $g_{33,m}^{*} = \max g_{33}^{*}(m, \rho, x)|_{\rho; x=\text{const}} = g_{33}^{*}(m_3, \rho, x)$ undergoes small changes at $0.10 \leq x \leq 0.30$ and $\rho = \text{const}$. It is obvious that the presence of prolate FC inclusions in the 0–3 composite favours an increase in the g_{33}^{*} values in comparison with the monolithic FC.

As follows from Table 5.1, $\max\left(d_{33}^{(1)}/\left|d_{31}^{(1)}\right|\right)$ is observed near the molar concentration $x = 0.25$. The volume-fraction dependences of the normalised piezoelectric coefficient $g_{33}^{*}/g_{33}^{(1)}$ of the 0–3 $(Pb_{0.75}Ca_{0.25})TiO_3$ FC/polyurethane composite (curves 1–3 in Fig. 5.3) suggest that a significant increase in the piezoelectric sensitivity is observed at aspect ratios $\rho \ll 1$ and volume fractions $0 < m \leq 0.20$. Increasing ρ and/or m results in a reduction of the g_{33}^{*} value (see, e.g., curve 4 in Fig. 5.3) due to a decrease of d_{33}^{*} or an increase of $\varepsilon_{33}^{*\sigma}$ on increasing m.

The calculated $g_{33}^{*}/g_{33}^{(1)}$ ratios, see curves 1–4 in Fig. 5.3, are comparable with experimental values (see sets 5 and 6 in Fig. 5.3) for the related 0–3 composites [8, 39]. The calculated curves 1–3 in Fig. 5.3 and values of $g_{33,m}^{*}$ demonstrate high piezoelectric sensitivity for the 0–3 composite with highly prolate spheroidal inclusions made from the highly anisotropic FC. Large $g_{33}^{*}/g_{33}^{(1)}$ values, curves 1 and 2 in Fig. 5.3, are achieved even in the presence of the piezo-passive polyurethane matrix. It should be noted for comparison that the experimental $g_{33}^{*}/g_{33}^{(1)}$ values shown as sets 5 and 6 in Fig. 5.3 are related to 0–3 FC/polymer composites with two piezoelectric components. Thus, the large piezoelectric anisotropy $\left(d_{33}^{(1)}/\left|d_{31}^{(1)}\right|\right) \gg 1$ of the FC component and the system of aligned prolate spheroidal FC inclusions in the 0–3 composite favour its high piezoelectric sensitivity. It is probable that the piezoelectric sensitivity of such a composite depends on the piezoelectric matrix to a lesser degree.

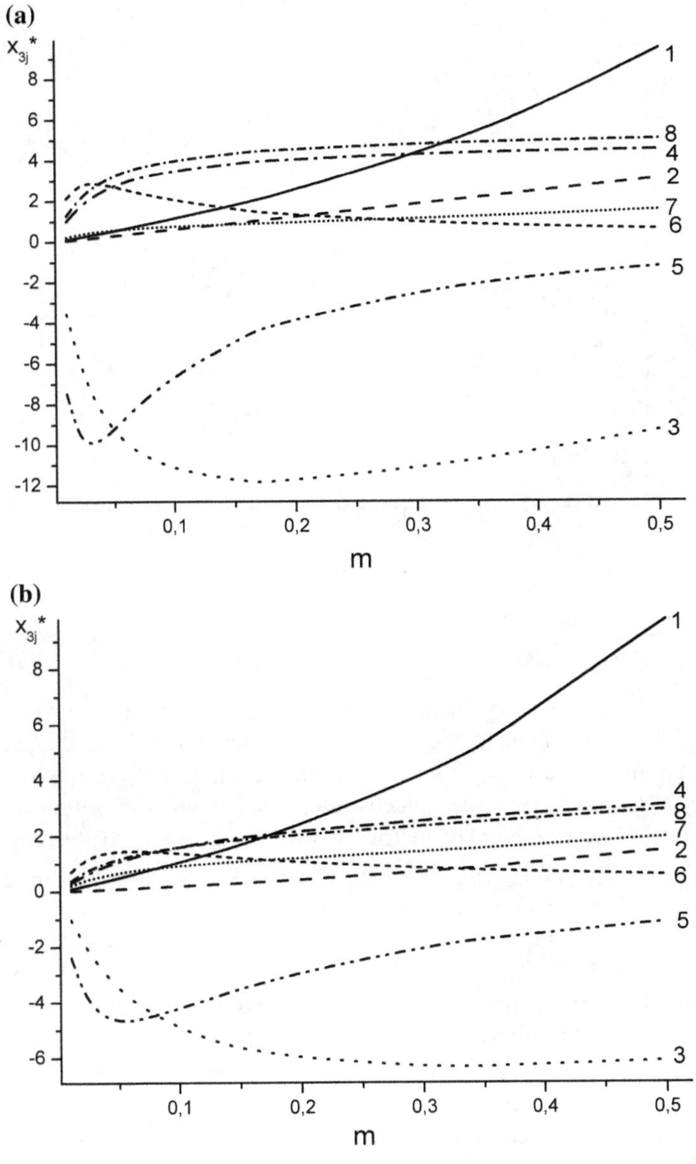

Fig. 5.2 Effective piezoelectric coefficients $e_{3j}^*(m, \rho)$ (in C/m^2), $d_{3j}^*(m, \rho)$ (in pC/N), $g_{3j}^*(m, \rho)$ (in mV m/N), and $h_{3j}^*(m, \rho)$ (in 10^8 V/m) which have been calculated for the 0–3 $(Pb_{0.9625}La_{0.025})(Ti_{0.99}Mn_{0.01})O_3$ FC/araldite composite containing spheroidal inclusions with the aspect ratio $\rho = 0.01$ (**a**), $\rho = 0.10$ (**b**) or $\rho = 1$ (**c**). The schematic of the 0–3 composite is shown in Fig. 5.1. Curve 1 corresponds to $10^2\ e_{31}^*$, curve 2 corresponds to e_{33}^* (graphs **a** and **b**) or $10^2\ e_{33}^*$ (graph **c**), curve 3 corresponds to d_{31}^* (graphs **a** and **b**) or 10 d_{31}^* (graph **c**), curve 4 corresponds to $10^{-1}\ d_{33}^*$ (graphs **a** and **b**) or d_{33}^* (graph **c**), curve 5 corresponds to $10^{-1}\ g_{31}^*$ (graphs **a** and **b**) or g_{31}^* (graph **c**), curve 6 corresponds to $10^{-2}\ g_{33}^*$ (graphs **a** and **b**) or $10^{-1}\ g_{33}^*$ (graph **c**), curve 7 corresponds to h_{31}^*, and curve 8 corresponds to $10^{-1}\ h_{33}^*$ (graphs **a** and **b**) or h_{33}^* (graph **c**) (reprinted from paper by Glushanin et al. [18], with permission from Elsevier)

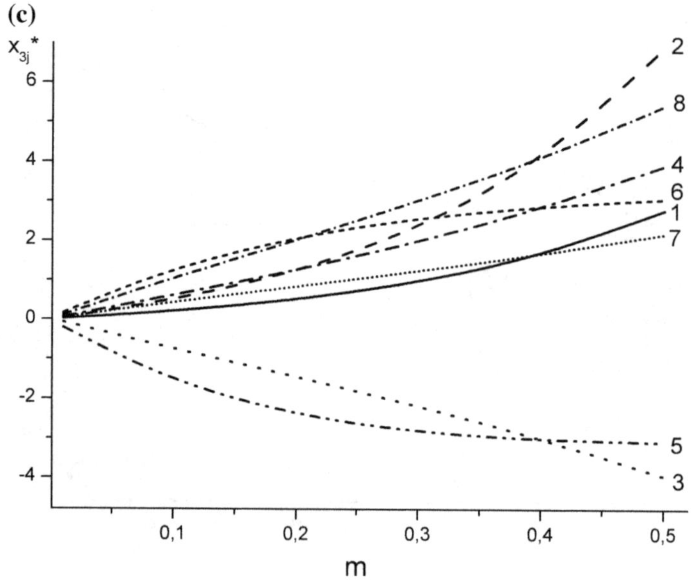

Fig. 5.2 (continued)

In the third example, we consider a link between the piezoelectric anisotropy $d_{33}^{(1)}/d_{31}^{(1)}$ of the FC component and the normalised piezoelectric coefficient $d_{33}^{*}/d_{33}^{(1)}$ of the 0–3 composite based on this FC component (Fig. 5.4). As earlier, we consider the $PbTiO_3$-type FC as a main piezoelectric component of the 0–3 composite. Despite the large anisotropy of the piezoelectric coefficients $d_{3j}^{(1)}$ and positive signs of the piezoelectric coefficients $e_{31}^{(1)}$ and $e_{33}^{(1)}$ in the $PbTiO_3$-type FC [see (5.5) and examples in [23, 24] and Table 5.1], the $d_{33}^{*}(m)$ dependence in the 0–3 composite based on this FC material is analogous to that found for 0–3 composites based, for example, on the $BaTiO_3$ or Pb(Zr, Ti)O_3 FCs [14, 23]. This is a result of the influence of the longitudinal piezoelectric response in the presence of the system of the isolated FC inclusions with the large piezoelectric coefficient $d_{33}^{(1)}$.

The largest $d_{33}^{*}/d_{33}^{(1)}$ values (curves 1 and 2 in Fig. 5.4a, c, and curve 5 in Fig. 5.4b) are related to $PbTiO_3$-based composites with highly prolate FC inclusions, i.e. at the aspect ratio $\rho \ll 1$. Comparing the normalised piezoelectric coefficient $d_{33}^{*}/d_{33}^{(1)}$ calculated for the $PbTiO_3$-type and Pb(Zr, Ti)O_3-type composites at $\rho = \text{const}$ and $\rho \ll 1$ (see Fig. 5.4a, b), we note that the lower $d_{33}^{*}/d_{33}^{(1)}$ ratios are achieved in the 0–3 composites based on the Pb(Zr, Ti)O_3-type FCs. Such a weakening of the longitudinal piezoelectric activity is a result of the relatively small piezoelectric anisotropy $d_{33}^{(1)}/d_{31}^{(1)}$ (see Table 1.2) and the equality $\text{sgn}\, e_{33}^{(1)} = -\text{sgn}\, e_{31}^{(1)}$ in the Pb(Zr, Ti)O_3-type FCs [23, 32]. It is seen from Fig. 5.4c that the condition $d_{33}^{*}/d_{33}^{(1)} \approx 1$ holds in the presence of the highly prolate $PbTiO_3$-type FC inclusions (at $\rho \ll 1$) and relatively soft polymer. It should be added that the trend $d_{33}^{*}/d_{33}^{(1)} \to 1$ is often

Table 5.1 Calculated values of room-temperature elastic moduli $c_{ab}^{(1),E}$ (in 10^{10} Pa),[a] piezoelectric coefficients $e_{ij}^{(1)}$ (in C/m^2), relative dielectric permittivities $\varepsilon_{pp}^{(1),\xi}/\varepsilon_0$, piezoelectric coefficient $g_{33}^{(1)}$ (in mV m/N) and piezoelectric anisotropy $d_{33}^{(1)}/d_{31}^{(1)}$ of the $(Pb_{1-x}Ca_x)TiO_3$ FC[b] (reprinted from paper by Glushanin et al. [18], with permission from Elsevier)

x	$c_{11}^{(1),E}$	$c_{12}^{(1),E}$	$c_{13}^{(1),E}$	$c_{33}^{(1),E}$	$c_{44}^{(1),E}$	$e_{31}^{(1)}$
0.10	19.1	5.22	5.15	18.6	6.85	0.676
0.15	18.8	5.09	5.01	18.3	6.76	0.751
0.20	18.5	4.76	4.68	18.0	6.79	0.844
0.23	18.7	5.03	4.93	18.2	6.76	1.06
0.24	18.4	4.72	4.61	17.9	6.76	1.07
0.25	18.9	5.19	5.06	18.3	6.76	1.33
0.26	18.9	5.14	5.02	18.3	6.77	1.36
0.30	18.6	4.89	4.80	18.1	6.77	1.08
x	$e_{33}^{(1)}$	$e_{15}^{(1)}$	$\varepsilon_{11}^{(1),\xi}/\varepsilon_0$	$\varepsilon_{33}^{(1),\xi}/\varepsilon_0$	$g_{33}^{(1)}$	$d_{33}^{(1)}/d_{31}^{(1)}$
0.10	3.90	1.62	111	109	21.1	−11.7
0.15	4.15	1.73	118	115	21.4	−13.0
0.20	4.31	1.77	126	123	20.6	−18.7
0.23	4.55	1.80	137	133	19.7	−30.3
0.24	4.69	1.87	145	141	19.4	−39.8
0.25	5.09	1.95	152	147	19.5	−71.9
0.26	5.30	2.03	161	156	19.1	−65.2
0.30	5.48	2.27	202	197	20.2	−17.6

[a]For the poled FC with ∞mm symmetry, the elastic modulus $c_{66}^{(1)}$ obeys the condition $c_{66}^{(1)} = \left(c_{11}^{(1),E} - c_{12}^{(1),E}\right)/2$ [32, 40]

[b]The full sets of electromechanical constants of the poled FC have been calculated by the self-consistent (effective medium) method [49, 50]. Spherical grains of the FC sample are assumed to be split into 90° lamellar domains of two types with equal volume fractions, and these domains are separated by planar 90° domain walls. These domain walls are assumed to be practically motionless so that a contribution from their displacements under external electric or mechanical fields [51] into the electromechanical constants of the FC approaches zero [49, 50]. The electromechanical constants of the single-domain crystal sample have been evaluated by the method from work [49] on the basis of experimental room-temperature data on the $(Pb_{1-x}Ca_x)TiO_3$ solid solutions

observed in the widespread 1–3 FC/polymer composites [3, 12–14, 22, 23] where the main piezoelectric component is either the Pb(Zr, Ti)O_3-type FC or $PbTiO_3$-type FC.

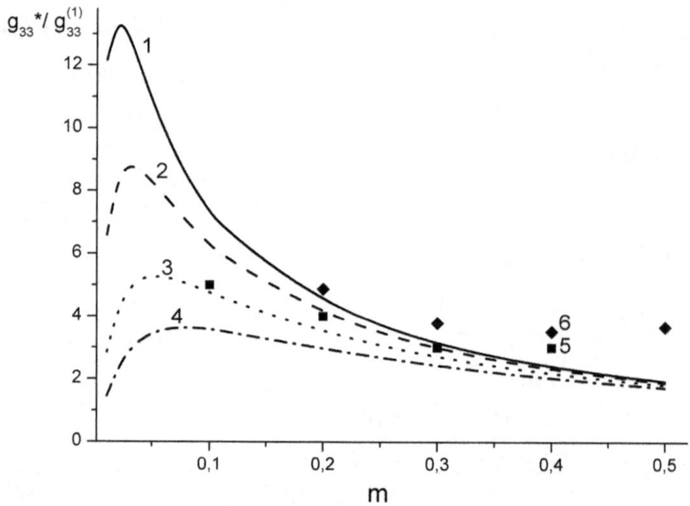

Fig. 5.3 Calculated (curves 1–4) and experimental (sets 5 and 6) volume fraction dependences of the $g_{33}^*/g_{33}^{(1)}$ ratio of the 0–3 FC/polymer composites (see the schematic in Fig. 5.1). Curves 1, 2, 3 and 4 are plotted for the $(Pb_{0.75}Ca_{0.25})TiO_3$ FC/polyurethane composite at the aspect ratio $\rho = 0.01$, 0.05, 0.10 and 0.15, respectively, experimental dependences 5 and 6 are related to the $PbTiO_3$ FC/70/30 mol% copolymer of vinylidene fluoride and trifluoroethylene composite [8] and (Pb, Ca)TiO_3 FC/75/25 mol% copolymer of vinylidene fluoride and trifluoroethylene composite [39], respectively (reprinted from paper by Glushanin et al. [18], with permission from Elsevier)

5.1.3 0–3 Composites Based on Pb(Zr, Ti)O_3-Type Ceramics

In Sect. 5.1.3, we consider the piezoelectric performance of the 0–3 FC/polymer composites where the FC component belongs to the Pb(Zr, Ti)O_3 type. As follows from Table 1.2 and numerous experimental data [32, 40], the piezoelectric anisotropy of the Pb(Zr, Ti)O_3-type FCs is small in comparison to the $PbTiO_3$-type FCs, and for the piezoelectric coefficients $e_{ij}^{(1)}$ of the Pb(Zr, Ti)O_3-type FCs conditions

$$e_{31}^{(1)} < 0, e_{33}^{(1)} > 0 \text{ and } e_{15}^{(1)} > 0 \tag{5.6}$$

hold instead of (5.5).

As seen from Fig. 5.5, the piezoelectric coefficients d_{3j}^* of the 0–3 composite follow a monotonic volume-fraction behaviour, and the piezoelectric coefficients g_{3j}^* of the same composite exhibit a non-monotonic behaviour irrespective of the aspect ratio ρ of the FC inclusion. Such a distinction between d_{3j}^* and g_{3j}^* is caused by the important role of the dielectric permittivity $\varepsilon_{33}^{*\sigma}$ in forming the piezoelectric response of the composite. We remind the reader that for the studied 0–3 FC/polymer composites poled along the OX_3 axis, the relation

$$g_{3j}^* = d_{3j}^*/\varepsilon_{33}^{*\sigma} \tag{5.7}$$

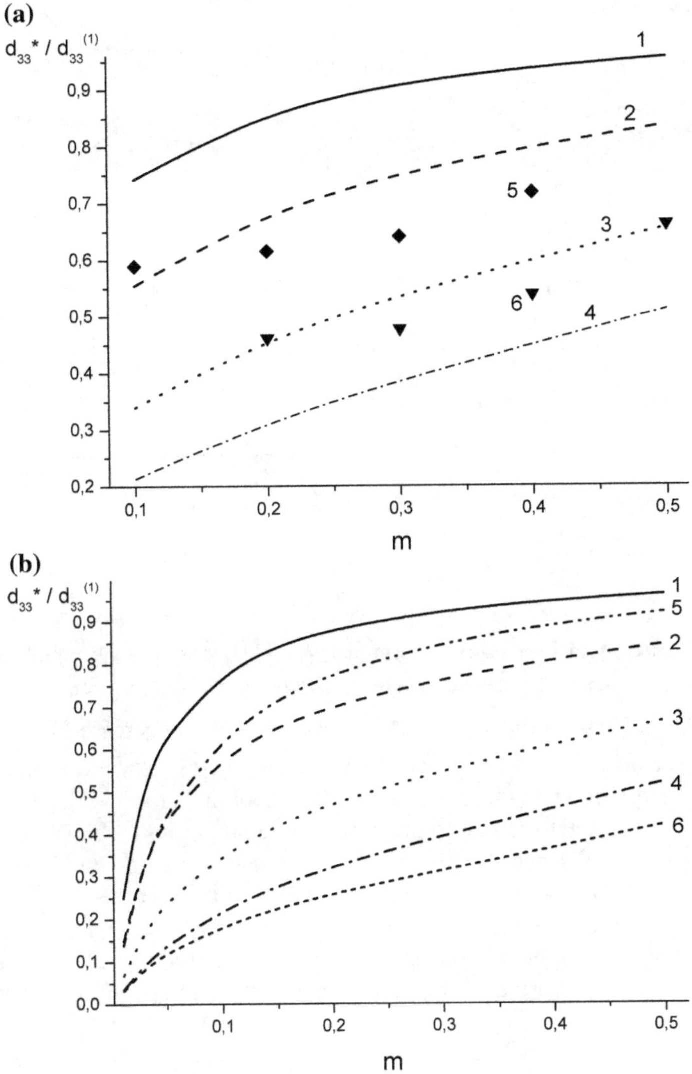

Fig. 5.4 Volume-fraction dependences of the normalised piezoelectric coefficient $d_{33}^{*}/d_{33}^{(1)}$ of the 0–3 FC/polymer composites (see the schematic in Fig. 5.1): **a** curves 1, 2, 3 and 4 are calculated for the $(Pb_{0.9625}La_{0.025})(Ti_{0.99}Mn_{0.01})O_3$ FC/araldite composite with the aspect ratio $\rho = 0.01$, 0.05, 0.10 and 0.15, respectively; experimental dependences 5 and 6 are related to the $PbTiO_3$ FC/70/30 mol% copolymer of vinylidene fluoride and trifluoroethylene composite [8] and (Pb, Ca)TiO_3 FC/75/25 mol% copolymer of vinylidene fluoride and trifluoroethylene composite [39], respectively, **b** curves 1, 2, 3 and 4 are calculated for the $(Pb_{0.75}Ca_{0.25})TiO_3$ FC/araldite composite with the aspect ratio $\rho = 0.01$, 0.05, 0.10 and 0.15, respectively; curves 5 and 6 are calculated for the PZT FC /polymer composite with $\rho = 0.01$ and 0.10 [14], respectively, and **c** curves 1, 2, 3 and 4 are calculated for the $(Pb_{0.75}Ca_{0.25})TiO_3$ FC/polyurethane composite with the aspect ratio $\rho = 0.01$, 0.05, 0.10 and 0.15, respectively (reprinted from paper by Glushanin et al. [18], with permission from Elsevier)

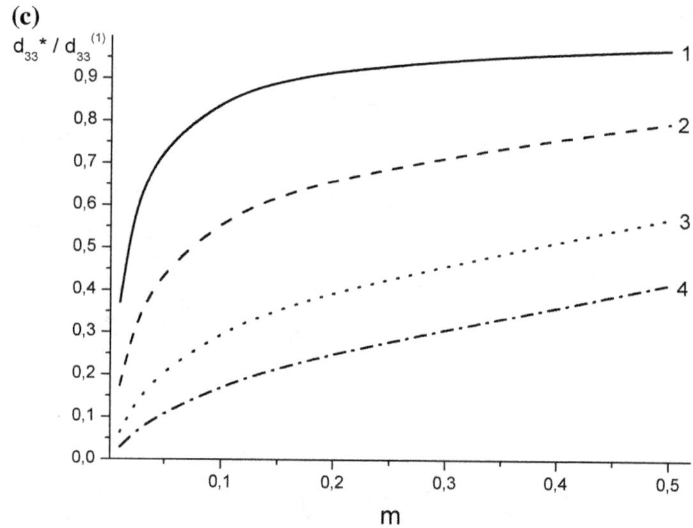

Fig. 5.4 (continued)

holds, and the effect of the combination [23, 24, 41] of the dielectric and piezoelectric properties leads to a non-monotonic $g_{3j}^*(m)$ dependence; see curves 3 and 4 in Fig. 5.5. An increase in the aspect ratio ρ leads to a rapid decrease of $\left|d_{3j}^*\right|$ at $m =$ const (cf., for instance, Fig. 5.5a, b). On increasing the aspect ratio ρ, the shape of the FC inclusion shown in Fig. 5.1 becomes less favourable for poling the composite sample along OX_3 and to achieve high piezoelectric properties in this 0–3 composite.

A comparison of graphs in Fig. 5.6 suggests that the aspect-ratio effect in the 0–3 composite is strong, even at relatively small volume fractions of FC $m \leq 0.15$. We observe the more intensive decrease of the piezoelectric coefficient $d_{33}^*(\rho)$ in comparison to $\left|d_{31}^*(\rho)\right|$ at $m =$ const and $0.01 < \rho < 0.1$ (*cf.* Fig. 5.6a, b). This is to be taken into account during the manufacture of similar FC/polymer composites.

5.1.4 0–3-Type Composites with Two Kinds of Ceramic Inclusions

Now we modify the 0–3 composite structure shown in Fig. 5.1 and assume that the matrix surrounding the spheroidal FC inclusions is also a 0–3 FC/polymer composite containing the same components. The spheroidal FC inclusions in the 0–3 matrix are distributed uniformly, and the remanent polarisation vector of each FC inclusion therein is oriented along the OX_3 axis, which is the poling axis for the whole composite sample. The shape of each FC inclusion in the 0–3 matrix is described by the equation $\left(x_1/a_{1,i}\right)^2 + \left(x_2/a_{1,i}\right)^2 + \left(x_3/a_{3,i}\right)^2 = 1$ in the rectangular co-ordinate sys-

Fig. 5.5 Volume-fraction dependences of the piezoelectric coefficients d_{3j}^{*} (in pC/N) and g_{3j}^{*} (in mV m/N) of the 0–3 PZT-5 FC/araldite composite at the aspect ratio $\rho = 0.01$ (**a**), 0.05 (**b**) and 0.10 (**c**). The schematic of the 0–3 composite is shown in Fig. 5.1

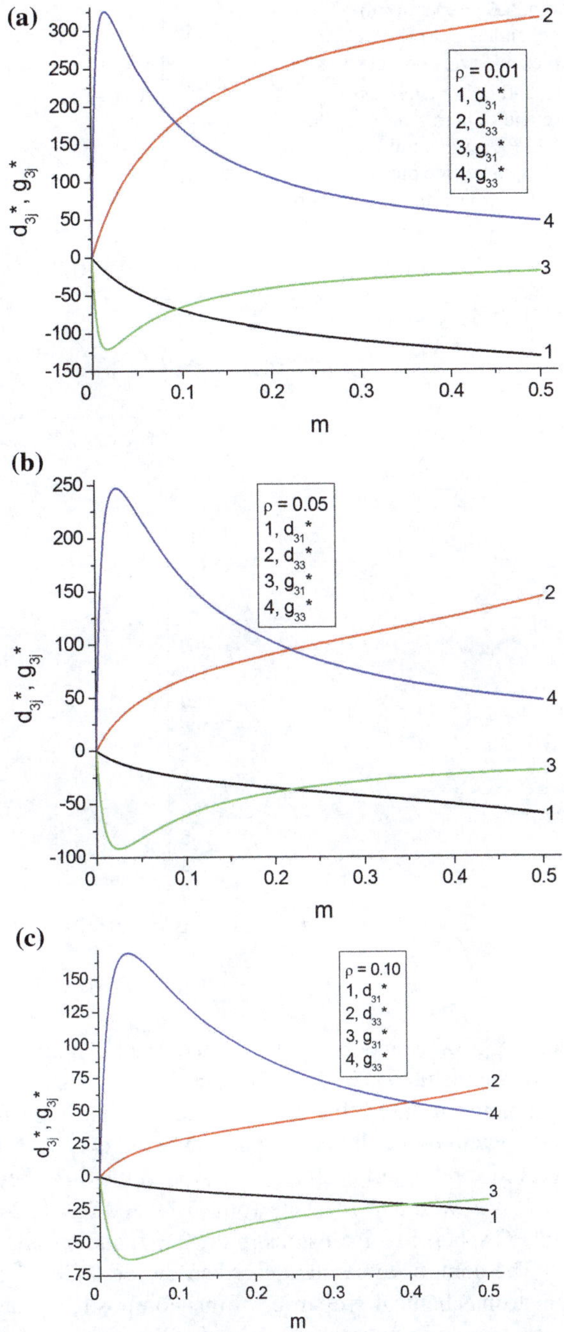

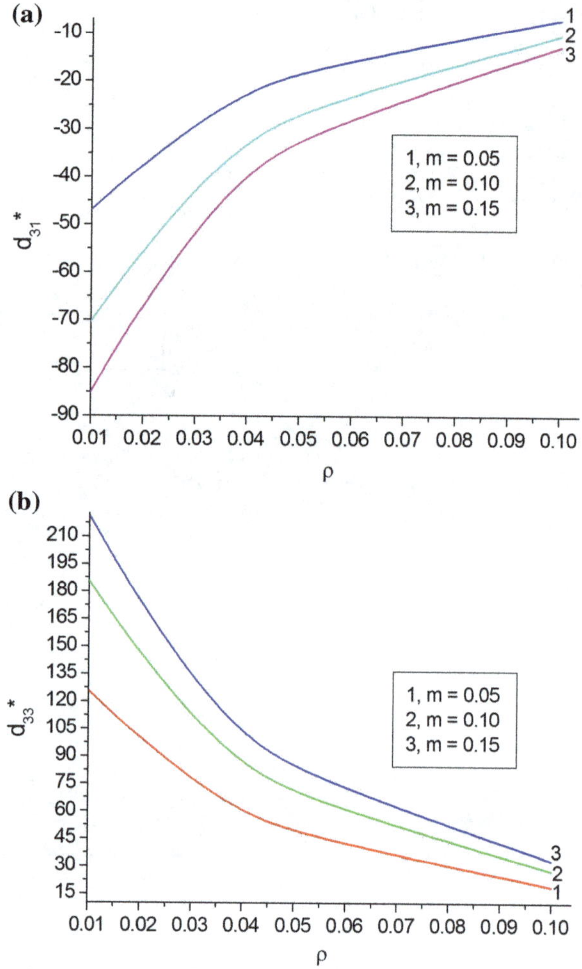

Fig. 5.6 Aspect-ratio dependences of the piezoelectric coefficients d_{3j}^* (**a** and **b**, in pC/N) and g_{3j}^* (**c** and **d**, in mV m/N) of the 0–3 PZT-5 FC/araldite composite (see the schematic in Fig. 5.1) at the aspect ratio $\rho \ll 1$

tem ($X_1X_2X_3$), and the aspect ratio of the FC inclusion is $\rho_i = a_{1,i}/a_{3,i}$. By analogy with the inclusions in the 0–3 FC/polymer composite (see Fig. 5.1b), the centres of symmetry of the FC inclusions in the 0–3 matrix occupy sites of a simple tetragonal lattice with unit-cell vectors parallel to the OX_j axes. We also assume that the semi-axes $a_{1,i}$ and $a_{3,i}$ of each FC inclusion in the 0–3 obey the conditions $a_{1,i} \ll a_j$ and $a_{3,i} \ll a_j$ where a_j is taken from (5.1), and $j = 1$, 2 and 3. The volume fraction of the FC spheroidal inclusions in the 0–3 matrix is m_i.

The composite can be described by the 0–3/0–3 connectivity, and the effective electromechanical properties of this composite are determined in two stages. In the first stage, we apply the effective field method and evaluate the effective properties of the 0–3 matrix in accordance with (5.3). Hereby elements of the $||C^*||$ matrix from (5.3) depend on m_i and ρ_i. In the second stage, we apply the same method and (5.3)

Fig. 5.6 (continued)

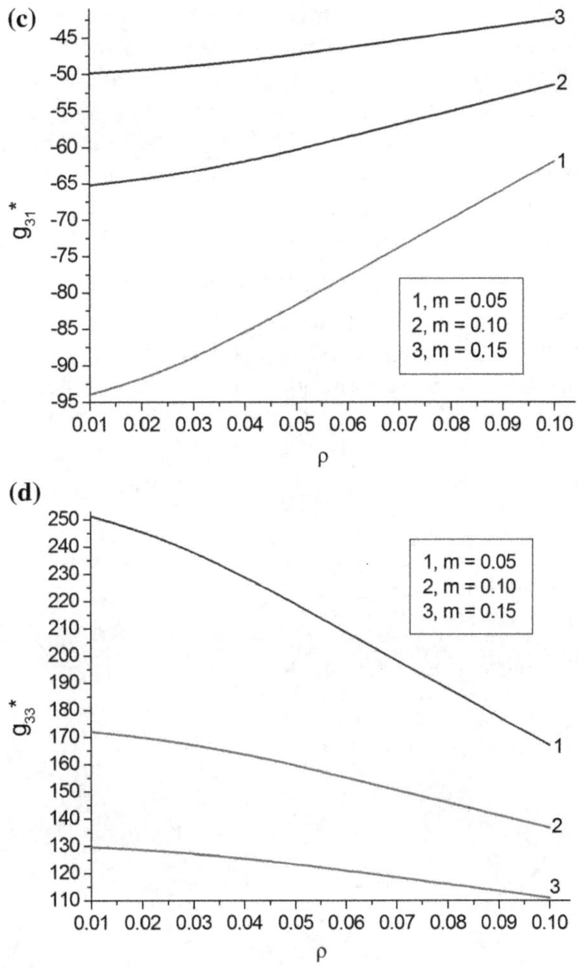

to calculate the effective properties of the 0–3-type composite as a whole. In this situation, the properties found in the first stage is related to the second component, i.e. the $||C^{(2)}||$ matrix in (5.3) would be equal to $||C^{*}||$ found in the first stage. Then the effective properties of the studied composite can be denoted as $X^{*} = X^{*}(m, \rho, m_i, \rho_i)$. Our calculations are performed by using the experimental properties of the FC and polymer components; see data in Tables 1.2 and 2.2.

Examples of the piezoelectric performance of the 0–3/0–3 composite based on the PZT-5 FC are shown in Fig. 5.7. It is seen that the longitudinal piezoelectric effect concerned with d_{33}^{*} (Fig. 5.7a) and g_{33}^{*} (Fig. 5.7c) is more sensitive to changes in the volume fraction m and aspect ratio ρ in comparison to the transverse piezoelectric effect (see Fig. 5.7b). This is accounted for by the role of the FC inclusions in the 0–3 matrix. Despite the relatively small volume fraction $m_i = 0.1$, these inclusions are in the form of the highly prolate spheroid at the aspect ratio $\rho_i = 0.1$ which results in anisotropic elastic properties in the 0–3 matrix, thereby increasing the piezoelectric anisotropy in the composite as a whole. Comparing Fig. 5.7a, b, we determine the m and ρ regions where the piezoelectric coefficients of the composite obey the condition $d_{33}^*/|d_{31}^*| = g_{33}^*/|g_{31}^*| \approx 5$. This means that the relatively large piezoelectric anisotropy of the composite is achieved in the presence of the PZT-5 FC that is the only piezoelectric component with a small anisotropy of the piezoelectric coefficients $d_{3j}^{(1)}$. As follows from Table 1.2, these piezoelectric coefficients obey the ratio $d_{33}^{(1)}/\left|d_{31}^{(1)}\right| = 2.19$. The local max g_{33}^* values at $\rho < 0.2$ (see Fig. 5.7c) are approximately equal to max g_{33}^* of a 1–3 relaxor-ferroelectric $0.7Pb(Mg_{1/3}Nb_{2/3})O_3 - 0.3PbTiO_3$ single-crystal/epoxy composite [7] with the domain-engineered single-crystal rods in the form of the rectangular parallelepiped. The values of the piezo-

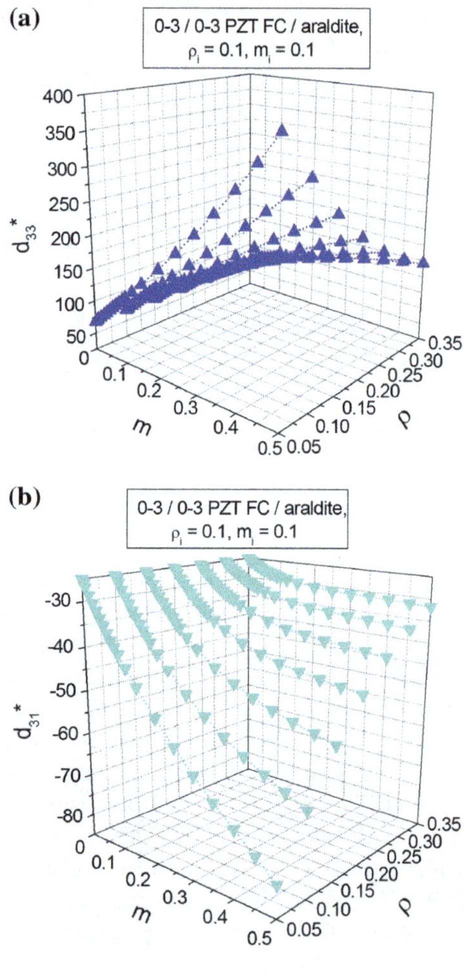

Fig. 5.7 Volume-fraction (m) and aspect-ratio (ρ) dependences of the piezoelectric coefficients d_{3j}^* (**a** and **b**, in pC/N) and g_{33}^* (**c**, in mV m/N) of the 0–3/0–3 PZT-5 FC/araldite/PZT-5 FC/araldite composite at the volume fraction $m_i = 0.1$ and the aspect ratio $\rho_i = 0.1$ in the 0–3 matrix

Fig. 5.7 (continued)

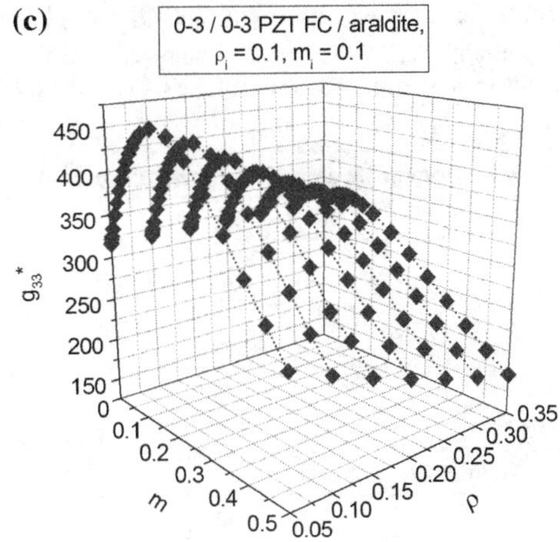

electric coefficient g_{33}^* of the 0–3 composite at $\rho = 0.2$ and 0.3 (see Table 5.2) are comparable to g_{33}^* from experimental data [30] on a DEP-structured composite.

Now we continue our comparison of the piezoelectric performance of 0–3-type composites based on the PZT-5 FC. We consider Table 5.2 where data on the piezoelectric coefficients d_{3j}^* and g_{33}^* are shown for two connectivity patterns. It is seen that the 0–3 FC/polymer matrix with heavily prolate inclusions ($\rho_i = 0.1$) promotes a stronger piezoelectric response from the composite as a whole. A similar 0–3 composite at $\rho = 0.1$ and $m = \text{const}$ is characterised, for example, by the piezoelectric coefficient d_{33}^* that becomes about 3–5 times smaller than d_{33}^* of the 0–3/0–3 composite; see Table 5.2. A similar decrease is also observed for the piezoelectric coefficient g_{33}^* of the 0–3 composite. The anisotropy of the piezoelectric coefficients d_{3j}^* of the 0–3 composite is smaller than that of the 0–3/0–3 composite in the studied volume-fraction (m) range. As follows from Table 5.2, increasing the aspect ratio ρ at $m = \text{const}$ in the 0–3 composite leads to a considerable decrease of its d_{33}^*, $|d_{31}^*|$ and g_{33}^*, and this is achieved due to the use of a piezo-passive araldite matrix that does not provide effective electromechanical coupling between the isolated piezoelectric FC inclusions (Fig. 5.1). Thus, the 0–3 matrix surrounding the prolate FC inclusions in the 0–3/0–3 composite promotes the better piezoelectric performance in comparison to the performance of the 0–3 composite based on the same FC.

5.2 1–3 Connectivity Patterns and Properties of Composites

The 1–3 FC/polymer composite shown in Fig. 5.8 is characterised by a continuous distribution of the FC component along one co-ordinate axis, which is often parallel to the poling axis. In the same 1–3 composite, the polymer component is self-connected in three dimensions; see the composite matrix in Fig. 5.8. The FC component can be distributed as a continuous row of particles, as a long rod (fibre), or a series of

Table 5.2 Piezoelectric coefficients d_{3j}^{*} (in pC/N) and g_{33}^{*} (in mV m/N) of the 0–3/0–3 PZT-5 FC/araldite/PZT-5 FC/araldite composite and 0–3 PZT-5 FC/araldite composite with spheroidal inclusions (see the schematic of the 0–3 composite in Fig. 5.1)

m	d_{31}^{*}	d_{33}^{*}	g_{33}^{*}
0–3/0–3 composite at $m_i = 0.1$ and $\rho_i = \rho = 0.1$			
0.05	−32.5	95.0	390
0.10	−38.8	121	428
0.15	−44.1	147	439
0.20	−48.6	172	430
0.25	−52.5	198	408
0.30	−55.8	223	379
0–3 composite at $\rho = 0.1$			
0.05	−7.05	19.0	167
0.10	−10.4	27.7	137
0.15	−12.7	33.4	111
0.20	−14.7	38.0	92.7
0.25	−16.6	42.2	79.4
0.30	−18.4	46.4	69.3
0–3 composite at $\rho = 0.2$			
0.05	−2.26	6.16	92.0
0.10	−3.55	9.54	93.5
0.15	−4.53	12.0	84.7
0.20	−5.37	14.0	75.5
0.25	−6.18	15.8	67.5
0.30	−6.99	17.7	60.9
0–3 composite at $\rho = 0.3$			
0.05	−1.07	2.94	54.1
0.10	−1.80	4.85	64.4
0.15	−2.39	6.32	63.9
0.20	−2.91	7.58	60.6
0.25	−3.42	8.56	56.6
0.30	−3.93	9.94	52.7

Calculations were carried out within the framework of the effective field method

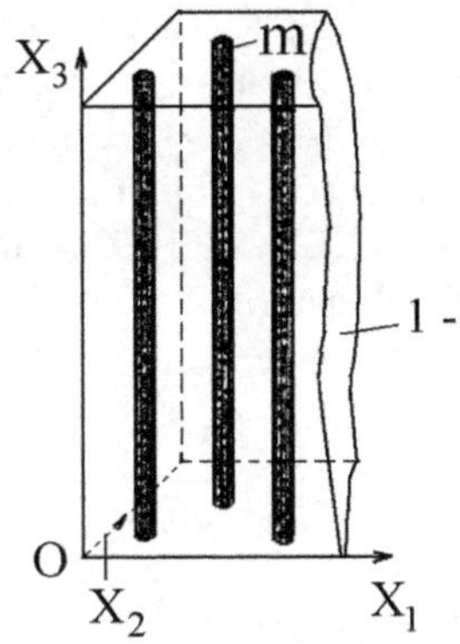

Fig. 5.8 Schematic of the 1–3 composite. m and $1 - m$ are volume fractions of the rods (component 1) and the matrix (component 2), and $(X_1X_2X_3)$ is the rectangular co-ordinate system (reprinted from monograph by Topolov and Bowen [23], with permission from Springer)

discs [5, 23, 26]. The manufacturing methods concerned with rapid prototyping [5, 28] are suitable to obtain composite samples with specific configurations of the rod bases. However in many cases, the FC rods are in the form of circular cylinders [2, 3, 23, 24] aligned parallel to the poling OX_3 axis (Fig. 5.8).

For the prediction of the effective properties of the 1–3 FC/polymer composite, it is assumed that in this composite the long FC rods (component with $n = 1$) are distributed regularly in a continuous polymer matrix (component with $n = 2$), see Fig. 5.8. The FC rods have the form of a circular cylinder with fixed radius. Centres of bases of the cylindrical rods occupy apices of squares in the (X_1OX_2) plane. The 1–3 composite as a whole is characterised by a remanent polarisation vector $\boldsymbol{P}_r^* \uparrow\uparrow OX_3$.

A full set of the effective electromechanical constants of the studied 1–3 FC/polymer composite can be found using different methods; see, for instance, work [3, 12, 15, 19–25, 32]. Hereby the effective field method is one of the main methods suitable for the prediction of the effective properties. Based on the effective field method and matrix (5.3), we find the effective properties of the 1–3 FC/polymer composite using the Eshelby tensor [33, 38] at the aspect ratio $\rho = 0$. The calculated effective properties are in agreement with the results obtained in terms of formulae [12]: a difference between the values at the volume fraction of FC m = const does not exceed 1%. Taking into account the square arrangement of the circular cylinders (Fig. 5.8) with equal radii, we should keep in mind the upper bound $m_u = \pi/4$ for the volume fraction of FC in the 1–3 composite.

The piezoelectric performance of the 1–3 composite based on the $PbTiO_3$-type FC is characterised [22] by the presence of $\min d_{31}^*(m)$ (curves 1–4 in Fig. 5.9a), $\min g_{31}^*(m)$ (Fig. 5.9b) and $\max g_{33}^*(m)$ (Fig. 5.9c) while the $d_{33}^*(m)$ dependence is monotonic; see curves 5–8 in Fig. 5.9a. The important feature of the piezoelectric sensitivity of the studied composite consists in minor differences between curves 1–4 in Fig. 5.9b, c, and this is observed at changes in the molar concentration x from $x = 0.10$ to $x = 0.30$. This is due the combined influence [23, 24, 41] of the dielectric and piezoelectric properties of the components. Taking into account this effect and validity of (5.7), we obtain $g_{31}^*(m) \approx$ const and $g_{33}^*(m) \approx$ const (see Fig. 5.9b, c).

In the PZT-5-based composite, the piezoelectric coefficients $d_{3j}^{*}(m)$ exhibit the monotonic behaviour (see curves 1 and 2 in Fig. 5.10); however, the piezoelectric coefficients $g_{3j}^{*}(m)$ are non-monotonic (see curves 3 and 4 in Fig. 5.10). The $g_{3j}^{*}(m)$ dependence is similar to that shown in Fig. 5.9b, c for the 1–3 composite based on the $(Pb_{1-x}Ca_x)TiO_3$ FC. The non-monotonic behaviour of g_{3j}^{*} is caused by the effect of combination of the dielectric and piezoelectric properties accordance with (5.7), while the extreme points of $g_{3j}^{*}(m)$ are related to $m \ll 1$, i.e. these points are in the

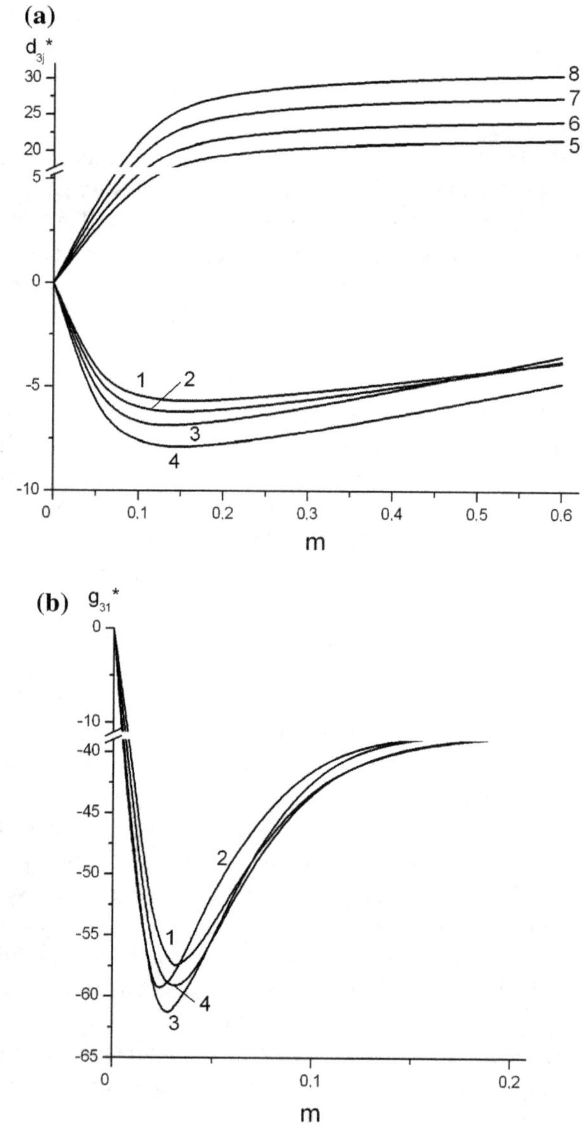

Fig. 5.9 Volume-fraction dependences of piezoelectric coefficients d_{3j}^{*} (**a**, in pC/N), g_{31}^{*} (**a**, in mV m/N) and g_{33}^{*} (**b**, in mV m/N) of the 1–3 $(Pb_{1-x}Ca_x)TiO_3$ FC/araldite composite (see the schematic in Fig. 5.7). In **a**, curves 1–4 correspond to $j = 1$, and curves 5–8 correspond to $j = 3$. Curves 1 and 5 are related to the composite with $x = 0.10$, curves 2 and 6 are related to the composite with $x = 0.20$, curves 3 and 7 are related to the composite with $x = 0.25$, and curves 4 and 8 are related to the composite with $x = 0.30$. In **b** and **c**, curves 1, 2, 3 and 4 are related to the composite with $x = 0.10$, 0.20, 0.25 and 0.30, respectively (reprinted from paper by Topolov et al. [22], with permission from IOP Publishing)

Fig. 5.9 (continued)

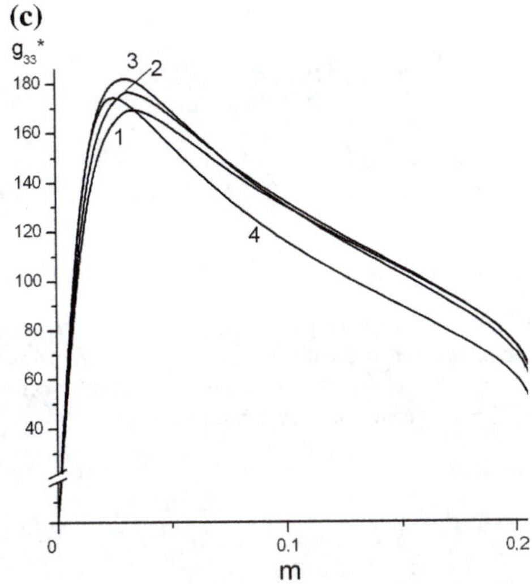

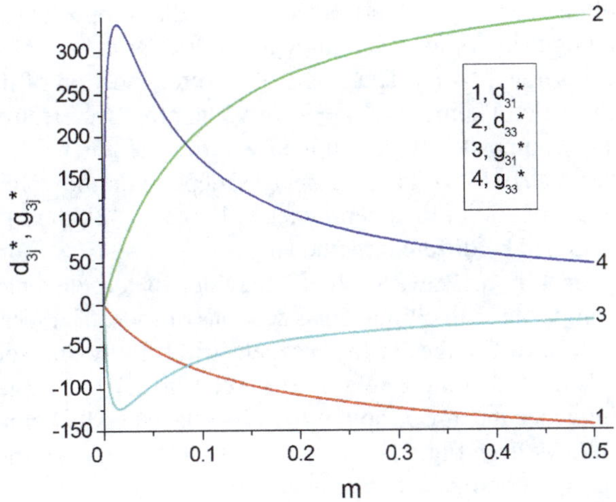

Fig. 5.10 Volume-fraction dependences of piezoelectric coefficients d_{3j}^* (in pC/N) and g_{3j}^* (in mV m/N) of the 1–3 PZT-5 FC/araldite composite (see the schematic in Fig. 5.8)

volume-fraction range where the piezoelectric coefficients $d_{3j}^*(m)$ demonstrate rapid changes; see Fig. 5.10.

The next important example of the piezoelectric performance is related to the 1–3 FC/polymer composite with elliptic cross sections of rods (Fig. 5.11) in the (X_1OX_2) plane that is perpendicular to the poling axis OX_3. As earlier, we assume that such a 1–3 composite is characterised by the regular distribution of FC rods in the large polymer matrix, i.e. centres of symmetry of the rod bases form a simple rectangular lattice in the (X_1OX_2) plane.

Cross sections of the FC rods by the (X_1OX_2) plane of the rectangular co-ordinate system $(X_1X_2X_3)$ shown in Fig. 5.11 are described by the equation

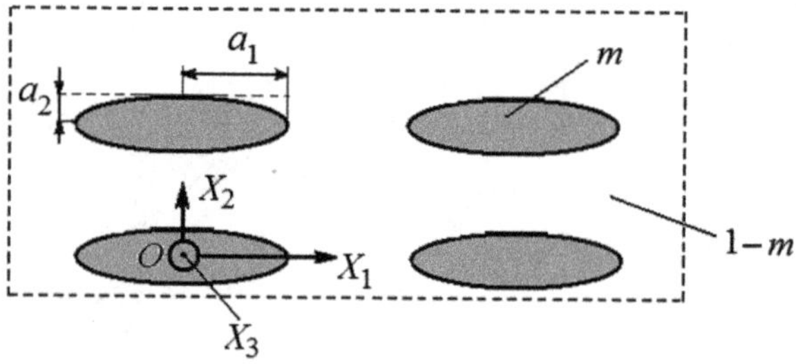

Fig. 5.11 Cross section of the 1–3 FC/polymer composite by the X_1OX_2 plane. ($X_1X_2X_3$) is the rectangular co-ordinate system, a_1 and a_2 are semi-axes of the ellipse, m is the volume fraction of FC, and $1 - m$ is the volume fraction of polymer (reprinted from paper by Topolov and Bisegna [42], with permission from Springer)

$$(x_1/a_1)^2 + (x_2/a_2)^2 = 1 \tag{5.8}$$

relative to the axes of the rectangular co-ordinate system ($X_1X_2X_3$). Semi-axes of the ellipse a_f from (5.8) are constant over the composite sample, the FC rods are aligned along the OX_3 axis as is shown in Fig. 5.8, and the height of each FC rod obeys the condition $h \gg a_f$. The shape of the cross section of the FC rod is characterised by the ratio of semi-axes $\eta = a_2/a_1$. The effective electromechanical properties of the 1–3 composite are determined within the framework of the effective field method; see matrix (5.3). These properties depend on the volume fraction of the FC rods m and the ratio of their semi-axes η. In work [42], the FEM is used to compare results obtained by different methods.

In Fig. 5.12, we show a behaviour of the piezoelectric coefficients of the 1–3 composite with elliptic cross sections of rods. Changes in the volume fraction of FC m lead to changes in the piezoelectric activity and sensitivity by analogy with the related 1–3 composite at $\eta = 1$, see Fig. 5.10. Changes in the ratio of semi-axes η influence the anisotropy of the piezoelectric effect along the OX_1 and OX_2 axes. As follows from Fig. 5.12, at $0 < \eta < 1$ and $m = \text{const}$, the piezoelectric coefficients of the 1–3 composite obey the conditions

$$\left|d_{32}^*\right| < \left|d_{31}^*\right| \text{ and } \left|g_{32}^*\right| < \left|g_{31}^*\right| \tag{5.9}$$

due to the orientation of elliptic cross sections [42] over the composite sample; see Fig. 5.11. We see that the extreme points of $g_{3j}^*(m)$ are observed at $m \ll 1$ irrespective of η, i.e. the important role of the dielectric permittivity $\varepsilon_{33}^{*\sigma}$ in forming the piezoelectric sensitivity in accordance with (5.7) is stated again. We note that the high level of the longitudinal piezoelectric sensitivity [42, 43] is achieved: for instance, the condition $g_{33}^* > 100$ mV m/N is achieved in the whole η range (Fig. 5.12f). This is to be taken into account in piezoelectric sensor and related applications.

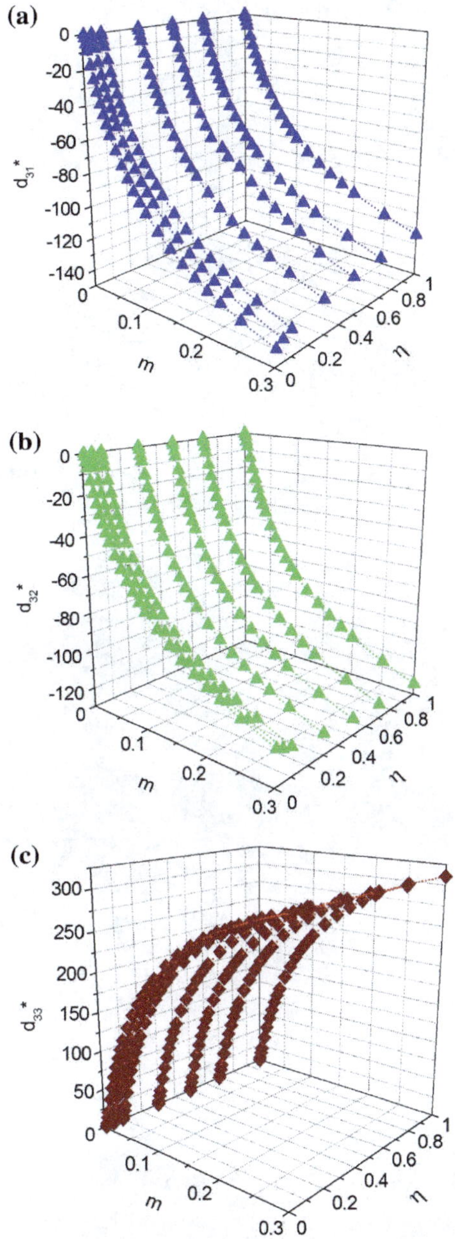

Fig. 5.12 Piezoelectric coefficients $d_{3j}^*(m, \eta)$ (**a–c**, in pC/N) and $g_{3j}^*(m, \eta)$ (**d–f**, in mV m/N) of the 1–3 PZT-5 FC/araldite composite. Cross section of the composite is shown in Fig. 5.11

Fig. 5.12 (continued)

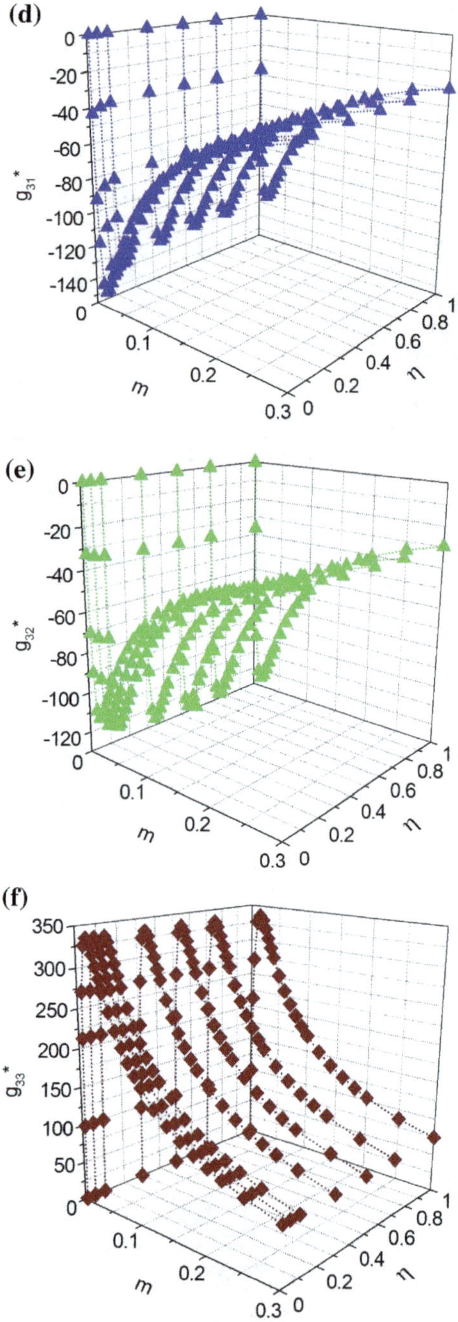

5.3 2–2 Connectivity and Properties of Parallel-Connected Composites

The 1–3 composite cross section shown in Fig. 5.11 suggests that in the limiting case of $\eta = 0$, the connectivity pattern undergoes changes, and the 2–2 connectivity is achieved. This is related to the parallel-connected composite [23, 24] with layers that are arranged regularly along the OX_2 axis and distributed continuously along the OX_1 and OX_3 axes. The poling axis of the composite is parallel to OX_3.

Graphs in Fig. 5.13 and data in Table 5.3 show that the transition from the 1–3 to 2–2 connectivity pattern in the FC/polymer composite has the evolution character [44], and no jumps in the piezoelectric coefficients are observed. It should be noted that the effective piezoelectric coefficients of the 1–3 and 2–2 composites (Table 5.3) were found by means of the effective field method and matrix method [23, 24], respectively. It is seen that the longitudinal piezoelectric effect undergoes very minor changes at $m = \text{const}$ and $0 \le \eta \le 1$. Otherwise speaking, changes in the microgeometry concerned with the non-polar OX_1 and OX_2 axes (see Fig. 5.11) do not affect the piezoelectric response of the composite along its poling direction that is parallel to OX_3.

Some examples of the piezoelectric performance of the 2–2 FC/polymer composites were analysed in work [23, 24, 43]. Here we do not consider the case of the series connection of the composite layers and the poling direction oriented perpendicular to the interfaces between the layers. In this case, a low piezoelectric activity of the 2–2 composite is observed in the wide volume-fraction range [23].

5.4 Electromechanical Coupling

As described in Sect. 1.4.3, the effectiveness of the energy conversion in the piezoelectric medium can be characterised by a set of ECFs that depend on electromechanical properties of this medium, vibration mode, etc. In a case of the piezo-active composite, its properties and ECFs depend on volume fractions of components, microgeometry and so on.

In Fig. 5.14, we show the volume-fraction behaviour of the ECFs that are concerned with the following oscillation modes: transverse [k_{31}^*, see (1.24)]; longitudinal [k_{33}^*, see (1.23)]; thickness [k_t^*, see (1.28)]; and planar [k_p^*, see (1.29)]. We also show some changes in ECFs caused by changes in the aspect ratio of the FC inclusions ρ in the 0–3 FC/polymer composite (Fig. 5.14a–c) and by the transition from 0–3 to 1–3 connectivity (cf. Fig. 5.14a, d). Validity of the condition

$$k_{33}^* \approx k_t^* \tag{5.10}$$

in the wide m range (see curves 2 and 4 in graphs of Fig. 5.14a–c) means that the electromechanical coupling of the 0–3 composite (Fig. 5.1) at $\rho \ll 1$ is similar to

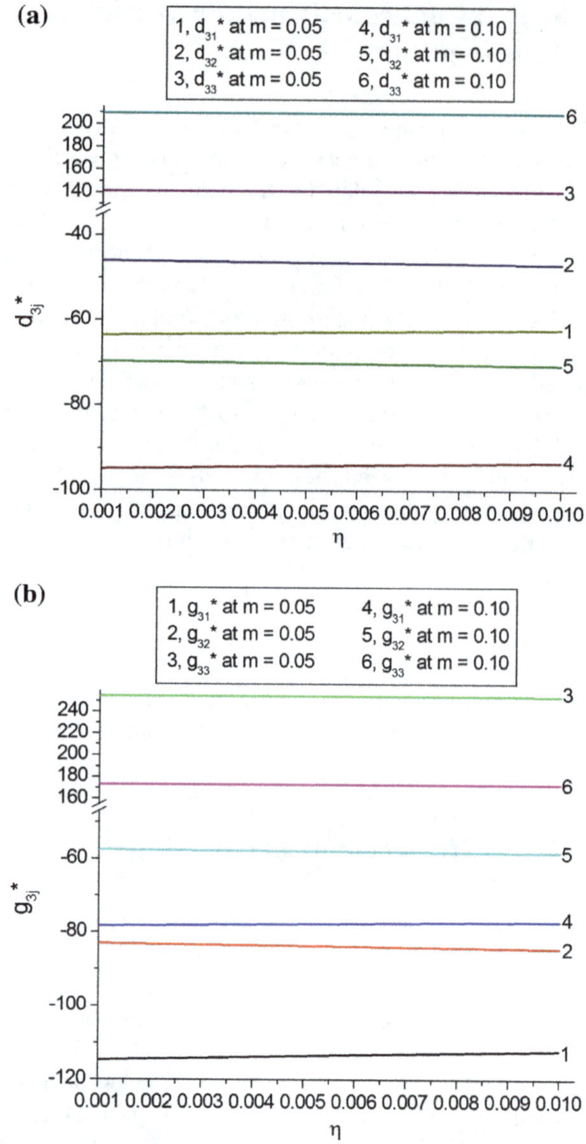

Fig. 5.13 Piezoelectric coefficients d_{3j}^* (**a**, in pC/N) and g_{3j}^*(**b**, in mV m/N) of the 1–3 PZT-5 FC/araldite composite at $m = \text{const}$ and $0.001 \le \eta \le 0.010$. Cross section of the composite is shown in Fig. 5.11

that of the 1–3 composite (Fig. 5.8). As follows from experimental data [3], the condition (5.10) remains correct for the 1–3 FC/polymer composite with cylindrical rods. It is also seen that the larger k_{33}^* value at the volume fraction of FC $m = \text{const}$ is achieved at the smaller aspect ratio ρ, and this is inseparably linked with the larger piezoelectric coefficient d_{33}^* at the smaller ρ value (see, for instance, Figs. 5.5 and 5.6b). A difference between k_{31}^* and k_p^* (see curves 1 and 3 in graphs of Fig. 5.14) is caused by elastic properties and their dependence on both m and ρ.

Table 5.3 Piezoelectric coefficients d^*_{3j} (in pC/N) and g^*_{3j}(in mV m/N) of the 1–3 and 2–2 PZT-5 FC/araldite composites

η	m	d^*_{31}	d^*_{32}	d^*_{33}	g^*_{31}	g^*_{32}	g^*_{33}
1–3 composite							
0.001	0.01	−17.5	−12.6	38.9	−147	−106	328
	0.03	−44.1	−31.8	98.0	−138	−99.2	305
	0.05	−63.6	−46.0	141	−115	−83.1	254
	0.10	−94.9	− 69.8	210	−78.4	−57.6	173
	0.20	−126	−95.9	277	−47.4	−36.1	104
	0.50	−157	−133	343	−21.6	−18.3	47.3
0.005	0.01	−17.2	−12.7	38.9	−145	−107	328
	0.03	−43.6	−32.1	98.0	−136	−100	305
	0.05	−62.9	−46.4	141	−113	−83.7	254
	0.10	−94.2	−70.2	210	−77.8	−57.9	173
	0.20	−125	−96.2	277	−47.2	−36.2	104
	0.50	−156	−133	343	−21.5	−18.3	47.3
2–2 composite							
0	0.01	−17.5	−12.6	38.9	−148	−105	328
	0.03	−44.3	−31.7	98.0	−138	−98.9	305
	0.05	−63.8	−45.9	141	−115	−82.9	254
	0.10	−95.1	−69.7	210	−78.5	−57.5	173
	0.20	−126	−95.8	277	−47.4	−36.0	104
	0.50	−157	−133	343	−21.6	−18.3	47.3

5.5 Piezoelectric Properties and Their Anisotropy

In Sect. 5.4, we considered four types of the piezoelectric coefficients from (1.12) to (1.15) and examples of the anisotropy concerned with the piezoelectric properties of the studied FC/polymer composites. We use data [45] obtained by means of the FEM. The FEM modelling was performed by using experimental data on the FC (Table 1.2) and polymer [3, 46] components. We remind the reader that the 1–3 and 0–3 composites are schematically shown in Figs. 5.1 and 5.8, respectively. As follows from Table 5.4, changes in the aspect ratio of the FC inclusion ρ influence all the types of the piezoelectric coefficients to a large extent. Relations between the piezoelectric coefficients d^*_{3j} and g^*_{3j} are caused by the effect of the combination of the piezoelectric and dielectric properties [1, 23, 24]. A combination of the piezoelectric and elastic properties leads to a large anisotropy of the piezoelectric coefficients e^*_{3j}: we state validity of the condition

$$e^*_{33}/\left|e^*_{31}\right| \gg 1 \tag{5.11}$$

at various ρ and m, see Table 5.4. Another example of the combination of the piezoelectric and dielectric properties is concerned with links between e^*_{3j} and h^*_{3j}. Moreover, the piezoelectric coefficients h^*_{3j} obey the condition

$$h^*_{33}/|h^*_{31}| \gg 1. \quad (5.12)$$

Inequality (5.12) is similar to (5.11) due to the relation [23, 24, 32] $h^*_{3j} = e^*_{3j}/\varepsilon^{*\xi}_{33}$ where $\varepsilon^{*\xi}_{33} = \varepsilon^{*\sigma}_{33} - 2d^*_{31}e^*_{31} - d^*_{33}e^*_{33}$ is the dielectric permittivity at mechanical strain $\xi = \text{const}$, and $\varepsilon^{*\sigma}_{33}$ is the dielectric permittivity at mechanical stress $\sigma = \text{const}$.

The similar behaviour of the piezoelectric coefficients of the 1–3 and 0–3 composites is observed (Table 5.5) after replacing the araldite matrix with a softer polyurethane matrix. This softening leads to larger values of the piezoelectric coef-

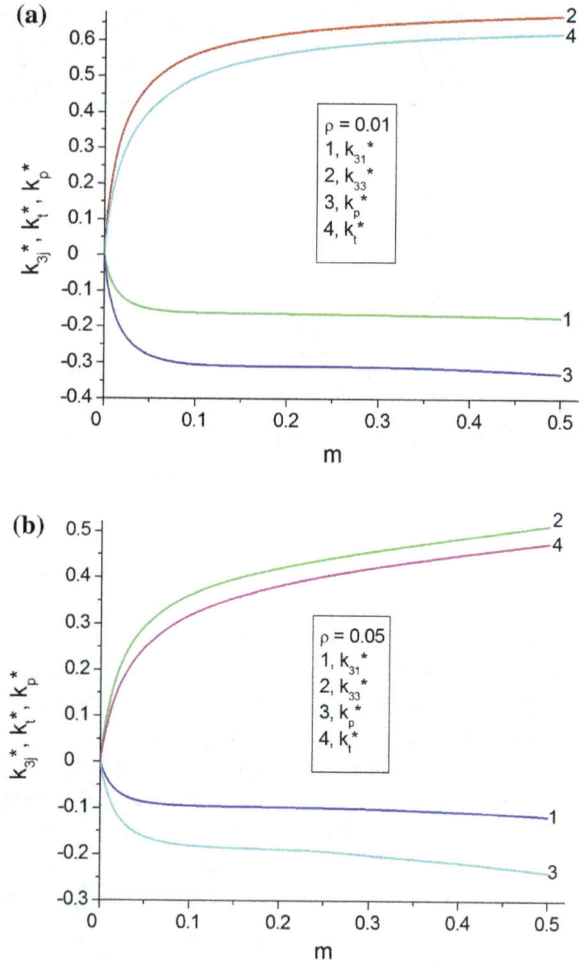

Fig. 5.14 Volume-fraction dependences of ECFs k^*_{3j}, k^*_p and k^*_t of the 0–3 PZT-5 FC/araldite composite (**a**–**c**) and 1–3 PZT-5 FC/araldite composite (**d**). Schematics of the 0–3 and 1–3 composites are shown in Figs. 5.1 and 5.8, respectively

Fig. 5.14 (continued)

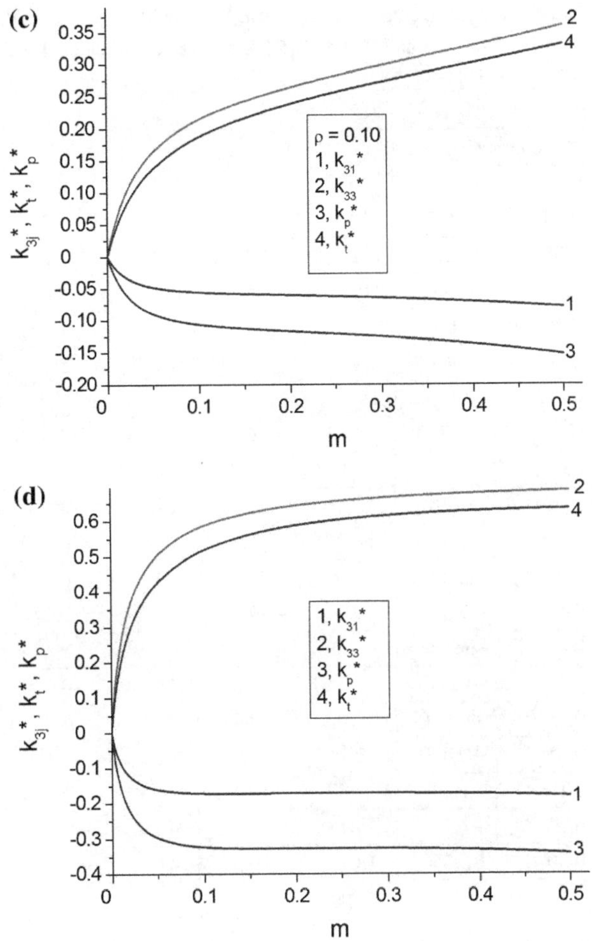

ficients d_{33}^* and g_{33}^* in comparison to data from Table 5.4: in this situation, the poled FC inclusions or rods undergo weaker mechanical stress fields caused by the surrounding matrix. It should be added that conditions (5.11) and (5.12) are also valid for the studied piezo-active polyurethane-containing composites.

An anisotropy concerned with the electromechanical properties of the composite poled along the OX_3 axis can be described in terms of the anisotropy factors [24]

$$\zeta_1^* = d_{33}^*/d_{31}^*,\ \zeta_2^* = k_{33}^*/k_{31}^* = \zeta_1^*\left(s_{11}^{*E}/s_{33}^{*E}\right)^{1/2} \text{ and } \zeta_3^* = k_t^*/k_p^*. \qquad (5.13)$$

Examples of the volume-fraction behaviour of the anisotropy factors from (5.13) are shown in Fig. 5.15. The graphs in Fig. 5.15 were built by taking into account the electromechanical properties found by means of the effective field method. Data in Fig. 5.15 suggest that the anisotropy factor ζ_2^* of the composite can be varied

Table 5.4 Piezoelectric coefficients d^*_{3j} (in pC/N), g^*_{3j} (in mV m/N), e^*_{3j} (in C/m^2) and h^*_{3j} (in GV/m) of the 1–3 and 0–3 PZT-5 FC/araldite composites, FEM data [45] (with permission from Prof. Dr. P. Bisegna)

m	d^*_{31}	d^*_{33}	g^*_{31}	g^*_{33}	$10^2\ e^*_{31}$	e^*_{33}	$10^2\ h^*_{31}$	h^*_{33}
1–3 composite, $\rho = 0$								
0.01	−14.4	39.0	−121	327	−0.426	0.196	−3.82	1.76
0.03	−36.5	98.2	−113	305	−1.30	0.589	−4.94	2.23
0.05	−52.8	141	−94.9	254	−2.21	0.981	−5.24	2.34
0.10	−79.6	210	−65.6	173	−4.64	1.96	−5.84	2.46
0.20	−108	277	−40.6	104	−10.3	3.91	−6.65	2.51
0.50	−144	343	−19.8	47.3	−39.7	9.67	−10.4	2.53
0–3 composite, $\rho = 0.1$								
0.01	−0.700	1.82	−16.4	42.6	−0.0540	0.00889	−1.27	0.208
0.03	−1.15	2.92	−22.9	58.1	−0.142	0.0169	−2.82	0.326
0.05	−1.38	3.46	−24.6	61.4	−0.224	0.0216	−3.99	0.384
0.10	−1.75	4.27	−25.0	60.9	−0.439	0.0333	−6.28	0.476
0.20	−2.33	5.62	−23.1	55.6	−0.999	0.0622	−9.92	0.618
0.50	−7.52	19.6	−17.3	45.2	−9.17	0.539	−21.7	1.28
0–3 composite, $\rho = 0.2$								
0.01	−0.290	0.750	−7.33	18.8	−0.0280	0.00351	−0.699	0.0875
0.03	−0.650	1.62	−13.9	34.7	−0.0870	0.00832	−1.87	0.178
0.05	−0.870	2.14	−16.8	41.3	−0.149	0.0121	−2.87	0.233
0.10	−1.24	2.97	−19.4	46.5	−0.320	0.0205	−5.01	0.321
0.20	−1.77	4.13	−19.7	46.0	−0.779	0.0397	−8.69	0.443
0.50	−5.25	13.1	−16.5	41.1	−7.00	0.312	−22.4	0.997

in a wider range in comparison to the anisotropy factors ζ^*_1 and ζ^*_3. This may be accounted for by the important role of the elastic compliances s^{*E}_{11} and s^{*E}_{33} in forming the anisotropy of the ECFs in accordance with (5.13). The s^{*E}_{11}/s^{*E}_{33} ratio undergoes considerable changes on increasing the volume fraction of FC m, and this influences the piezoelectric response and electromechanical coupling of the composite. The presence of the heavily prolate FC inclusions ($\rho \to 0$) in the composite, undoubtedly, promotes the considerable difference between the elastic compliances s^{*E}_{11} and s^{*E}_{33}.

The results on the piezoelectric properties and anisotropy factors of the studied composites are to be taken into consideration for piezoelectric sensor, transducer, energy-harvesting, and other applications wherein the longitudinal piezoelectric effect and related oscillation mode play the leading role.

Table 5.5 Piezoelectric coefficients d^*_{3j} (in pC/N), g^*_{3j} (in mV.m/N), e^*_{3j} (in C/m^2) and h^*_{3j} (in GV/m) of the 1–3 and 0–3 PZT-5 FC/polyurethane composites, FEM data [45] (with permission from Prof. Dr. P. Bisegna)

m	d^*_{31}	d^*_{33}	g^*_{31}	g^*_{33}	$10^2\ e^*_{31}$	e^*_{33}	$10^2\ h^*_{31}$	h^*_{33}
1–3 composite, $\rho = 0$								
0.01	−24.9	66.4	−207	552	−0.243	0.198	−2.27	1.84
0.03	−56.2	149	−161	427	−0.744	0.593	−2.87	2.28
0.05	−75.1	198	−123	325	−1.29	0.987	−3.07	2.40
0.10	−101	263	−76.8	200	−2.66	1.97	−3.36	2.49
0.20	−124	314	−44.2	112	−5.96	3.94	−3.84	2,54
0.50	−150	357	−20.3	48.2	−23.4	9.79	−6.12	2.56
0–3 composite, $\rho = 0.1$								
0.01	−0.780	2.01	−20.9	53.8	−0.0290	0.00550	−0.764	0.147
0.03	−1.22	3.05	−27.6	69.3	−0.0780	0.00900	−1.76	0.224
0.05	−1.42	3.50	−28.7	70.9	−0.124	0.0129	−2.51	0.261
0.10	−1.71	4.14	−27.9	67.4	−0.245	0.0196	−3.99	0.319
0.20	−2.10	5.26	−24.7	59.3	−0.566	0.0366	−6.39	0,414
0.50	−6.72	17.8	−17.5	46.5	−6.11	0.355	−16.2	0.944
0–3 composite, $\rho = 0.2$								
0.01	−0.300	0.750	−8.47	21.5	−0.0140	0.00192	−0.388	0.0547
0.03	−0.640	1.59	−15.8	39.0	−0.0440	0.00458	−1.08	0.112
0.05	−0.860	2.09	−18.9	45.9	−0.0777	0.00668	−1.69	0.147
0.10	−1.19	2.83	−21.3	50.4	−0.170	0.0114	−3.03	0.203
0.20	−1.64	3.81	−20.9	48.4	−0.424	0.0221	−5.40	0.282
0.50	−4.68	11.8	−16.7	42.0	−4.54	0.192	−16.4	0.694

5.6 Figures of Merit

In Sect. 1.4.3, we considered squared figures of merit concerned with the longitudinal piezoelectric effect, transverse piezoelectric effect and hydrostatic piezoelectric activity of a material; see (1.33)–(1.35). The squared figure of merit $(Q^*_{33})^2$ is related to the longitudinal piezoelectric effect in the composite and can be represented in the form

$$(Q^*_{33})^2 = (d^*_{33})^2/\varepsilon^{*\sigma}_{33}. \tag{5.14}$$

As follows from numerous data [23–25, 43], $(Q^*_{33})^2$ from (5.14) has a maximum point that can be achieved at various volume fractions of FC—from a few per cent to a few tens of per cent. This maximum is caused by a relatively rapid increase of the piezoelectric coefficient d^*_{33} (in the presence of the heavily prolate FC inclusions

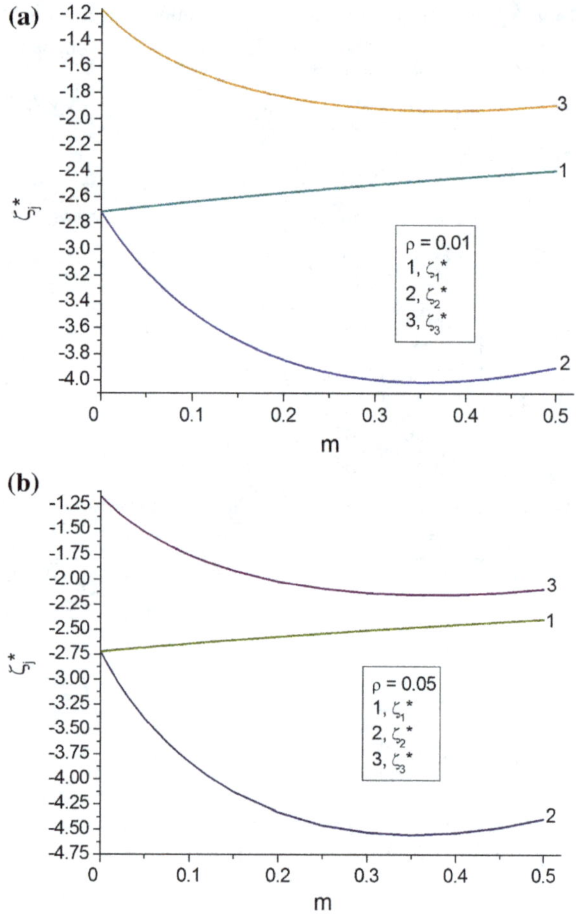

Fig. 5.15 Volume-fraction dependences of anisotropy factors ζ_j^* of the 0–3 PZT-5 FC/araldite composite (**a**–**c**) and 1–3 PZT-5 FC/araldite composite (**d**). Schematics of the 0–3 and 1–3 composites are shown in Figs. 5.1 and 5.8, respectively

or rods, see Figs. 5.1 and 5.8) and by the slow increase of the dielectric permittivity $\varepsilon_{33}^{*\sigma}$.

Some results on the squared figure of merit $\left(Q_{33}^*\right)^2$ of the studied composites are shown in Table 5.6. We state the drastic increase of this parameter of the composite at the transition from the 1–3 connectivity pattern at the aspect ratio $\rho = 0$ (see column 2 in Table 5.6) to the 0–3 connectivity pattern at the aspect ratio $\rho > 0$ (see column 3 in Table 5.6). This increase is concerned with the considerable weakening of the longitudinal piezoelectric effect at forming the isolated aligned prolate FC inclusions (see Fig. 5.1) instead of the continuous FC rods (see Fig. 5.8) that are parallel to the poling axis OX_3. In the presence of the polyurethane matrix that is softer than the araldite one, we achieve larger values of $\left(Q_{33}^*\right)^2$ at the same m and ρ parameters of the composite; see comparisons in Table 5.6. The larger $\left(Q_{33}^*\right)^2$ value is achieved mainly due to the larger piezoelectric coefficient d_{33}^*; see Table 5.5. It should be

Fig. 5.15 (continued)

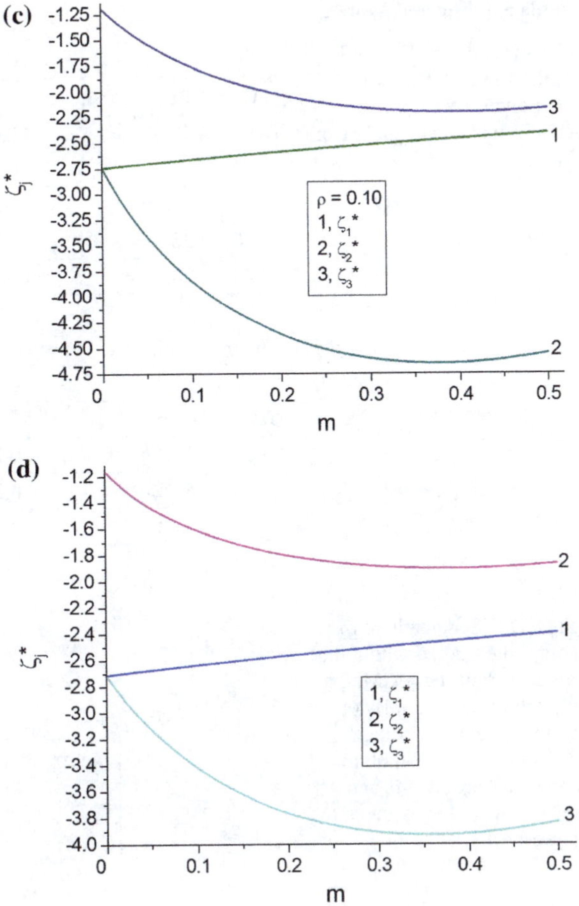

noted that our results on the squared figures of merit are to be taken into account in piezoelectric energy-harvesting, sensor, hydroacoustic and related applications.

5.7 Mixed Connectivity Model and Effective Properties

The MCM can be used to consider features of properties of a composite with 0–3 and 1–3 connectivity patterns [31, 39, 47]. This model is important to estimate the role of the 1–3 connectivity in the DEP alignment of the FC particles in the polymer matrix [48] and, therefore, in determining the effective properties of the composite. In the MCM, it is assumed that both parallel and series connectivity patterns coexist in the composite. The representative unit cell shown in Fig. 5.16 contains two structural elements: the first is in the form of a rectangular parallelepiped, and the second is in

Table 5.6 Squared figure of merit $(Q_{33}^{*})^2$ (in 10^{-12} Pa^{-1}) of the 1–3 and 0–3 PZT-5 FC/polymer composites (evaluated by using data from Tables 5.4 and 5.5)

m	$(Q_{33}^{*})^2$ at $\rho = 0$	$(Q_{33}^{*})^2$ at $\rho =$ 0.1	$(Q_{33}^{*})^2$ at $\rho =$ 0.2
FC/araldite composites			
0.01	10.8	0.0775	0.0141
0.03	30.0	0.170	0.0562
0.05	35.8	0.212	0.0884
0.10	36.3	0.260	0.138
0.20	28.8	0.312	0.190
0.50	16.2	0.886	0.538
FC/polyurethane composites			
0.01	36.7	0.108	0.0161
0.03	63.6	0.211	0.0620
0.05	64.4	0.248	0.0959
0.10	52.6	0.279	0.143
0.20	35.2	0.312	0.184
0.50	17.2	0.828	0.496

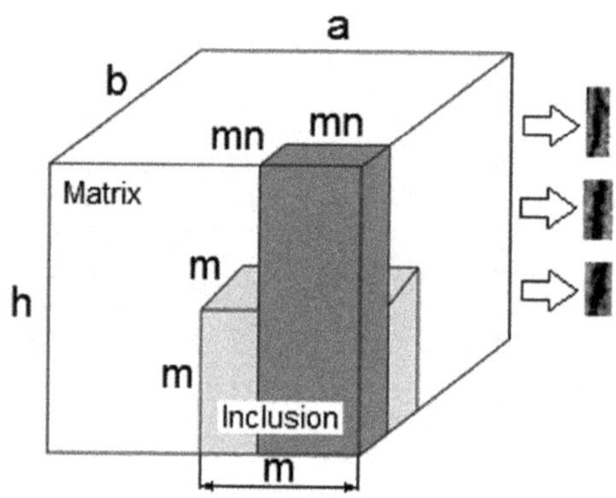

Fig. 5.16 Schematic of the composite studied within the framework of the MCM. On the right side, examples of similar fragments of the DEP-structured FC/polymer composite are shown, and the poling axis is assumed to be vertical

the form of a cube. Edges of the representative unit cell are $a = b = h = 1$ linear unit, and the distribution of FC in this unit cell is described in terms of parameters n and m. These parameters are related to the mixed connectivity of the composites (Fig. 5.16), namely 0–3 and 1–3. The FC component is represented as a cube of dimensions m where the nth fraction of this cube connects both electrodes oriented parallel to the horizontal plane. Rectangular co-ordinate axes OX_1, OX_2 and OX_3 are parallel to the edges a, b and h, respectively, and the composite is poled along the OX_3 axis, i.e. on the vertical direction.

The main MCM formulae related to the effective physical properties of the composite are given in work [31, 39, 47]. The effective properties and related parameters of the composite are represented as function of n and m. Taking into account Fig. 5.16, we determine that the volume fraction of the inclusion (or the FC component) in the composite is $m^3 + n^2m^2(1 - m)$.

In Fig. 5.17 we show calculations for a PT FC/epoxy composite within the framework of the MCM [47]. In general, the calculated piezoelectric and pyroelectric properties are in agreement with experimental data and comparable to data evaluated by using a variety of methods. In Fig. 5.17a, we observe a monotonic increase of the piezoelectric coefficient d_{33}^* with an increase of n at m = const. This indicates an improvement of the longitudinal piezoelectric activity of the composite is achieved by the formation of highly prolate FC inclusions (or long rods) along the poling axis OX_3, as is achieved by DEP structuring. Concerning the piezoelectric sensitivity of the composite, we note that its largest g_{33}^* value is achieved for a relatively small volume fraction of FC and at $n \to 1$ (see Fig. 5.17b), i.e. in the presence of the long FC rods aligned in the composite matrix parallel to the poling axis. This is consistent with results [23, 26, 43] on the 1–3 FC/polymer composites.

5.8 Conclusion

In this chapter, we have analysed examples of the piezoelectric properties and related parameters of 0–3, 1–3 and 2–2 FC/polymer composites, and 0–3-type composites with two kinds of the FC inclusions. The aforementioned composites are widespread and of interest because of their piezoelectric performance, electromechanical coupling, squared figures of merit, anisotropy of the properties and other characteristics. It has been shown that the microgeometry of the composite plays a key role in forming its piezoelectric sensitivity, and changes in the microgeometry can lead to appreciable changes in the piezoelectric sensitivity and related parameters.

The main examples of the piezoelectric sensitivity are concerned with composites that are based on either the PZT-type or $PbTiO_3$-type FCs. These FCs exhibit different electromechanical properties and different levels of piezoelectric anisotropy; see Table 1.2. The FC component and specifics of its distribution over the composite structure influence the anisotropy of the piezoelectric properties, ECFs from (5.13), and squared figures of merit. These and other factors are to be taken into account when manufacturing piezo-active composites with a variety of connectivity patterns.

The presence of the long FC inclusions, for example, rods or plates, oriented along the poling axis in the composite promotes an increase of its piezoelectric sensitivity. Of specific interest are non-monotonic volume-fraction dependences of the piezoelectric properties of the composites, especially the piezoelectric coefficients g_{3j}^*. Their maximum and minimum values are to be taken into consideration by the analysis of the piezoelectric sensitivity and methods to improve the parameter in the presence of specific connectivity patterns. It should be noted that values of $g_{3j}^* \sim$ 100 mV m/N (see Tables 5.4, 5.5, Fig. 5.17b, etc.) are achieved in specific volume-fraction and/or aspect-ratio ranges. The aforementioned g_{3j}^* values are approximately one order-of-magnitude larger than the piezoelectric coefficients $g_{3j}^{(1)}$ of poled FCs

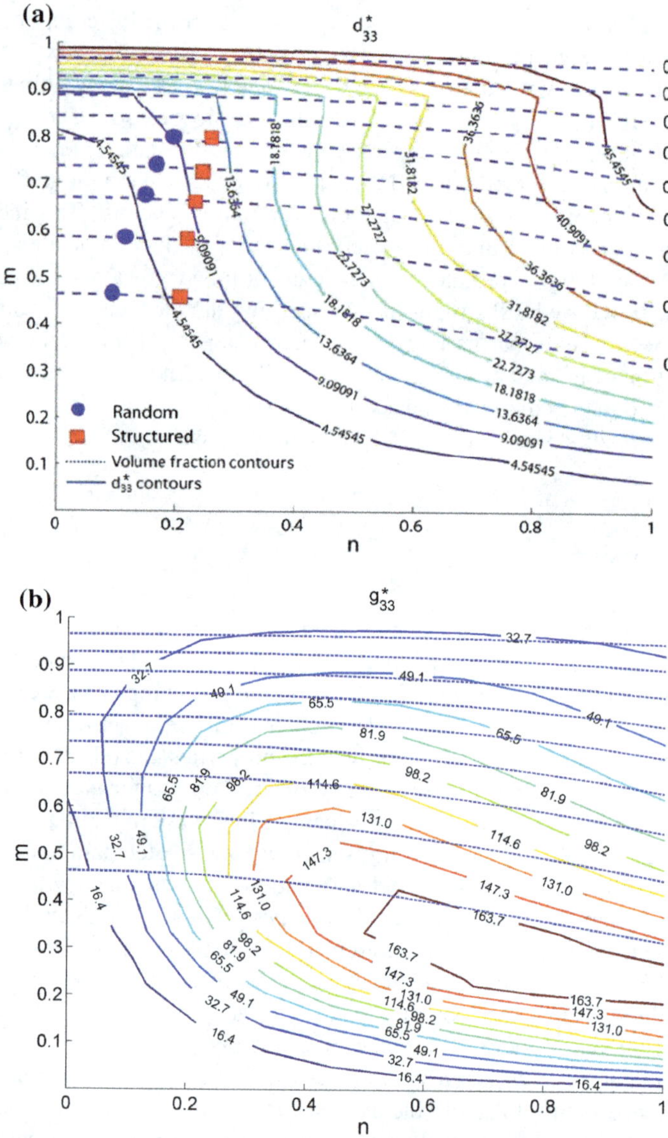

Fig. 5.17 Diagrams (contour charts) built for the 0–3/1–3 PT FC/epoxy composite by using the MCM: piezoelectric coefficients d_{33}^* (**a**, in pC/N) [30] and g_{33}^* (**b**, in mV m/N) [47], and pyroelectric coefficient p^* (**c**, in 10^{-6} C m^{-2} K^{-1}) [47]. Dashed lines are related to the constant volume fraction of FC (from bottom to top, 0.1, 0.2, … and 0.9) in the composite. Figure a has been reprinted from paper by Khanbareh et al. [30], with permission from IOP Publishing)

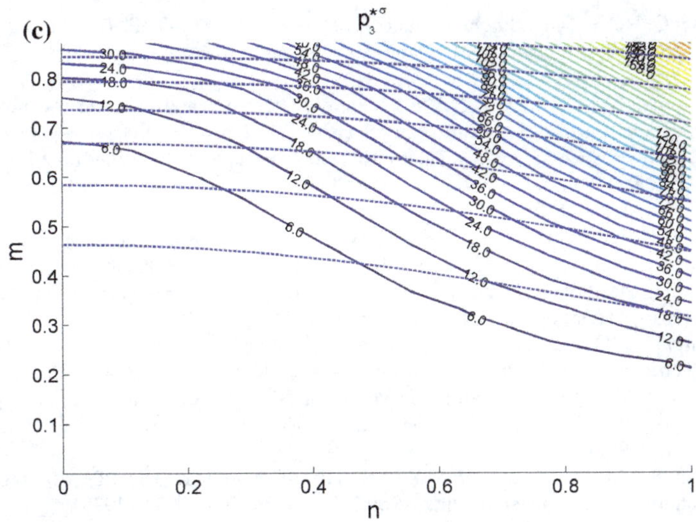

Fig. 5.17 (continued)

(see Table 1.3) as main components of the studied composites. The values of the piezoelectric coefficient $g_{33}^* \approx 100$ mV m/N at volume fractions of FC $m \geq 0.1$ are consistent with experimental results shown in Fig. 3.16c. We add that the studied piezoelectric coefficients g_{3j}^* are inseparably linked with the squared figures of merit $\left(Q_{3j}^*\right)^2$ of the composite, and large values of these parameters are to be taken into account in a range of piezoelectric sensor, energy-harvesting and other transducer applications. Data shown in Fig. 5.17c can be considered at the prediction of the pyroelectric properties of the PT/polymer composites.

References

1. R.E. Newnham, D.P. Skinner, L.E. Cross, Connectivity and piezoelectric-pyroelectric composites. Mater. Res. Bull. **13**, 525–536 (1978)
2. K.A. Klicker, J.V. Biggers, R.E. Newnham, Composites of PZT and epoxy for hydrostatic transducer applications. J. Am. Ceram. Soc. **64**, 5–9 (1981)
3. H.L.W. Chan, J. Unsworth, Simple model for piezoelectric ceramic/polymer 1–3 composites used in ultrasonic transducer applications. IEEE Trans. Ultrason. Ferroelectr. Freq. Control **36**, 434–441 (1989)
4. H.L.W. Chan, M.C. Cheung, C.L. Choy, Study on $BaTiO_3$/P(VDF–TrFE) 0–3 composites. Ferroelectrics **224**, 113–120 (1999)
5. E.K. Akdogan, M. Allahverdi, A. Safari, Piezoelectric composites for sensor and actuator applications. IEEE Trans. Ultrason. Ferroelectr. Freq. Control **52**, 746–775 (2005)
6. S.A. Wilson, G.M. Maistros, R.W. Whatmore, Structure modification of 0–3 piezoelectric ceramic/polymer composites through dielectrophoresis. J. Phys. D Appl. Phys. **38**, 175–182 (2005)

7. F. Wang, C. He, Y. Tang, Single crystal 0.7Pb(Mg1/3Nb2/3)O_3–0.3PbTiO_3 epoxy 1–3 piezoelectric composites prepared by the lamination technique. Mater. Chem. Phys. **105**, 273–277 (2007)
8. J.B. Ngoma, J.Y. Cavaille, J. Paletto, J. Perez, Dielectric and piezoelectric properties of copolymer-ferroelectric composite. Ferroelectrics **109**, 205–210 (1990)
9. H.L.W. Chan, Y. Chen, C.L. Choy, Thermal hysteresis in the permittivity and polarization of lead zirconate titanate/vinylidenefloride-trifluoroethylene 0–3 composites. IEEE Trans. Dielectr. Electr. Insul. **3**, 800–805 (1996)
10. Y. Hirata, T. Numazawa, H. Takada, Effects of aspect ratio of lead zirconate titanate on 1–3 piezoelectric composite properties. Jpn. J. Appl. Phys. Pt **1**(36), 6062–6064 (1997)
11. L. Pardo, J. Mendiola, C. Alemany, Theoretical treatment of ferroelectric composites using Monte Carlo calculations. J. Appl. Phys. **64**, 5092–5097 (1988)
12. A.A. Grekov, S.O. Kramarov, A.A. Kuprienko, Effective properties of a transversely isotropic piezoelectric composite with cylindrical inclusions. Mech. Compos. Mater. **25**, 54–61 (1989)
13. F. Levassort, M. Lethiecq, C. Millar, L. Pourcelot, Modeling of highly loaded 0–3 piezoelectric composites using a matrix method. IEEE Trans. Ultrason. Ferroelectr. Freq. Control **45**, 1497–1505 (1998)
14. V.M. Levin, M.I. Rakovskaja, W.S. Kreher, The effective thermoelectroelastic properties of microinhomogeneous materials. Int. J. Solids Struct. **36**, 2683–2705 (1999)
15. F. Levassort, V.Yu. Topolov, M. Lethiecq, A comparative study of different methods of evaluating effective electromechanical properties of 0–3 and 1–3 ceramic/polymer composites. J. Phys. D Appl. Phys. **33**, 2064–2068 (2000)
16. C.K. Wong, Y.M. Poon, F.G. Shin, Explicit formulas for effective piezoelectric coefficients of ferroelectric 0–3 composites based on effective medium theory. J. Appl. Phys. **93**, 487–496 (2003)
17. N. Fakri, L. Azrar, L. El Bakkali, Electroelastic behavior modeling of piezoelectric composite materials containing spatially oriented reinforcements. Int. J. Solids Struct. **40**, 361–384 (2003)
18. S.V. Glushanin, V.Yu. Topolov, A.V. Krivoruchko, Features of piezoelectric properties of 0–3 PbTiO_3-type ceramic/polymer composites. Mater. Chem. Phys. **97**, 357–364 (2006)
19. Yu.V. Sokolkin, A.A. Pan'kov, *Electroelasticity of Piezo-Composites with Irregular Structures* (Fizmatlit, Moscow, 2003) (in Russian)
20. R. Kar-Gupta, T.A. Venkatesh, Electromechanical response of 1–3 piezoelectric composites: effect of poling characteristics. J. Appl. Phys. **98**, 054102 (2005)
21. R. Kar-Gupta, T.A. Venkatesh, Electromechanical response of 1–3 piezoelectric composites: an analytical model. Acta Mater. **55**, 1093–1108 (2007)
22. V.Yu. Topolov, P. Bisegna, A.V. Krivoruchko, Features of electromechanical properties of 1–3 composites based on PbTiO_3-type ceramics. J. Phys. D Appl. Phys. **41**, 035406 (2008)
23. V.Yu. Topolov, C.R. Bowen, *Electromechanical Properties in Composites Based on Ferroelectrics* (Springer, London, 2009)
24. V.Yu. Topolov, P. Bisegna, C.R. Bowen, *Piezo-Active Composites. Orientation Effects and Anisotropy Factors* (Springer, Berlin, Heidelberg, 2014)
25. V.Yu. Topolov, C.R. Bowen, P. Bisegna, New aspect-ratio effect in three-component composites for piezoelectric sensor, hydrophone and energy-harvesting applications. Sens. Actuators A – Phys. **229**, 94–103 (2015)
26. T.R. Gururaja, A. Safari, R.E. Newnham, L.E. Cross, Piezoelectric ceramic/polymer composites for transducer applications, in *Electronic Ceramics: Properties, Devices, and Applications*, ed. by M. Levinson (Marcel Dekker, New York Basel, 1988), pp. 92–128
27. G.M. Garner, N.M. Shorrocks, R.W. Whatmore, M.T. Goosey, P. Seth, F.W. Ainger, 0–3 piezoelectric composites for large area hydrophones. Ferroelectrics **93**, 169–176 (1989)
28. A. Safari, E.K. Akdogan, Rapid prototyping of novel piezoelectric composites. Ferroelectrics **331**, 153–179 (2006)
29. C.A. Randall, D.V. Miller, J.H. Adair, A.S. Bhalla, Processing of electroceramic—polymer composites using the electrorheological effect. J. Mater. Res. **8**, 899–904 (1993)

30. H. Khanbareh, S. van der Zwaag, W. Groen, Effect of dielectrophoretic structuring on piezoelectric and pyroelectric properties of PT-epoxy composites. Smart Mater. Struct. **23**, 105030 (2014)
31. M.P. Wenger, D.K. Das-Gupta, Mixed connectivity composite material characterization for electroactive sensors. Polym. Eng. Sci. **39**, 1176–1188 (1999)
32. L.P. Khoroshun, B.P. Maslov, P.V. Leshchenko, *Prediction of Effective Properties of Piezo-Active Composite Materials* (Naukova Dumka, Kiev, 1989) (in Russian)
33. J.H. Huang, W.-S. Kuo, Micromechanics determination of the effective properties of piezoelectric composites containing spatially oriented short fibers. Acta Mater. **44**, 4889–4898 (1996)
34. T. Mori, K. Tanaka, Average stress in matrix and average elastic energy of materials with misfitting inclusions. Acta Metall. **21**, 571–574 (1973)
35. M.L. Dunn, M. Taya, An analysis of piezoelectric composite materials containing ellipsoidal inhomogeneities. Proc. R. Soc. (Lond.), Pt A **443**, 265–287 (1993)
36. M.L. Dunn, M. Taya, Micromechanics predictions of the effective electroelastic moduli of piezoelectric composites. Int. J. Solids Struct. **30**, 161–175 (1993)
37. C. Poizat, M. Sester, Homogénéisation périodique de composites piézoélectriques 0–3: influence de la distribution. Rev. des Compos. et des Matériaux Avancés **11**, 65–74 (2001)
38. J.H. Huang, S. Yu, Electroelastic Eshelby tensors for an ellipsoidal piezoelectric inclusion. Compos. Eng. **4**, 1169–1182 (1994)
39. C.J. Dias, D.K. Das-Gupta, Electroactive polymer-ceramic composites, in *Proceedings of the 4th International Conference on Properties and Applications of Dielectric Materials*, July 3–8, 1994, Brisbane, Australia (IEEE, Piscataway (1994), pp. 175–178
40. D.A. Berlincourt, D.R. Cerran, H. Jaffe, Piezoelectric and piezomagnetic materials and their function in transducers, in *Physical Acoustics. Principles and Methods*, Vol. 1: *Methods and Devices*, ed. Mason W (Pt A. Academic Press, New York London, 1964), pp. 169–270
41. R.E. Newnham, Nonmechanical properties of composites, in *Concise Encyclopedia of Composite Materials*, ed. by A. Kelly, R.W. Cahn, M.B. Bever (Elsevier, Oxford, 1994), pp. 214–220
42. V.Yu. Topolov, P. Bisegna, Anisotropic piezoelectric properties of 1–3 ceramic/polymer composites comprising rods with elliptic cross section. J. Electroceram. **25**, 26–37 (2010)
43. V.Yu. Topolov, C.R. Bowen, P. Bisegna, *Piezo-Active composites. Microgeometry – Sensitivity Relations* (Springer International Publishing Switzerland, 2018)
44. V.Yu. Topolov, S.V. Glushanin, Evolution of connectivity patterns and links between interfaces and piezoelectric properties of two-component composites. J. Phys. D Appl. Phys. **35**, 2008–2014 (2002)
45. P. Bisegna, Private communication (2018)
46. L.V. Gibiansky, S. Torquato, On the use of homogenization theory to design optimal piezocomposites for hydrophone applications. J. Mech. Phys. Solids **45**, 689–708 (1997)
47. H. Khanbareh, Expanding the functionality of piezo-particulate composites. Dissertation, Delft University of Technology, Delft, 2016
48. C. Dias, D. Das Gupta, Inorganic ceramic polymer ferroelectric composite electrets. IEEE Trans. Dielectr. Electr. Insul. **3**, 706–734 (1996)
49. V.Yu. Topolov, A.V. Turik, A.I. Chernobabov, On the mechanism of high piezoelectric anisotropy in lead titanate-based ferroelectrics. Crystallogr. Rep. **39**, 805–809 (1994)
50. V.Yu. Topolov, A.V. Turik, A.I. Chernobabov, On the piezoelectric anisotropy in modified $PbTiO_3$ ceramics. Ferroelectrics **154**, 271–276 (1994)
51. E.I. Bondarenko, V.Yu. Topolov, A.V. Turik, The role of 90° domain wall displacements in forming physical properties of perovskite ferroelectric ceramics. Ferroelectr. Lett. Sect. **13**, 13–19 (1991)

Chapter 6
From Microgeometry to Improved Properties of Piezo-Particulate Composites

Abstract The manufacture and performance of the structured piezo-particulate composites based on ferroelectric ceramics are considered by taking into account the important role of dielectrophoresis in forming the effective piezoelectric properties and related parameters of these composites. The considerable effect of topology, processing and physical properties of components on the effective properties of the studied composites is discussed to shed light on improving their piezoelectric properties. Some characteristics of related structured two- and three-component composites are compared. These composites can be used in piezo-technical, pyroelectric, energy-harvesting and other applications.

6.1 Dielectrophoresis and Structure – Properties Relations

In the present monograph, we have described the manufacture and effective properties of the DEP-structured piezo-particulate composites based on FCs. We have also shown the considerable effect of topology, processing and physical properties of components on the effective properties of these composites. In the focus of our attention are the DEP-structured two- and three-component composites whereby the main piezoelectric component chosen is based on the widespread perovskite-type FCs. Despite the piezoelectric passiveness of the polymer component, its chemical composition and physical properties should be taken into consideration for DEP-based composites because of many reasons, such as mechanical strength, temperature stability, electrical conductivity and so on. Traditionally, two kinds of the piezo-particulate composites have been the subject of our study, namely materials based on PT [1, 2] and materials based on PZT FCs [3, 4]. Of an independent interest are piezo-particulate composites suitable for high-temperature and pyroelectric applications.

An experimental study of various piezo-active composites and their effective properties is important in connection with the 'structure – properties' and 'composition – properties' sides of the fundamental triangle [5]. The use of DEP structuring improves the microgeometry of the piezo-particulate composites since a large number of FC particles form long chains in an external electric field $\boldsymbol{E}$ [1–4]. Improving the composite microgeometry by the formation of a system of long piezoelectric

H. Khanbareh et al., *Piezo-Particulate Composites*, Springer Series in Materials Science 283, https://doi.org/10.1007/978-3-030-19204-4_6

elements oriented parallel to the ***E*** vector promotes an improvement of the poling conditions for the composite system and facilitates an analysis of relations between the composite structure and effective properties therein.

A degree of alignment of the aforementioned multiparticle chain almost parallel to the ***E*** vector becomes a crucial microgeometric factor that leads to effective properties and further potential applications of the studied composites. Hereby, in addition to the traditional 0–3 connectivity, we have studied quasi 1–3 connectivity to compare the piezoelectric performance of the related piezo-active composites. The next important factor that strongly influences their performance is the presence of porosity in the polymer matrix. The formation of porous and foam structures in the polymer medium leads to an increase of the piezoelectric sensitivity of the composite [4] without a serious loss of its mechanical strength and flexibility.

DEP enables the manufacture of piezo-particulate PT-based composites with improved properties. As a rule, a pure PT FC is not stable in the mechanical sense due to large spontaneous strains and that requires chemical modification to be used in piezo-technical applications. The ability to surround the numerous PT particles with a polymer component leads to the formation of a composite structure and, therefore, avoids any cracking and destruction of the PT FC. The piezo-particulate PT/epoxy composites manufactured by means of DEP [2] are characterised by enhanced dielectric, piezoelectric and pyroelectric properties in comparison to conventional 0–3 composites [6] wherein a number of granular, unconnected and isolated particles are located in a continuous polymer matrix. The degree of structuring as well as the electric properties of the components has been the most important parameters that control the effective properties of the composite. The piezoelectric sensitivity associated with the piezoelectric coefficients g_{3j}^* [5, 7] is considered as a function of a number of variables, and ways to reach values of $g_{33}^* \sim 100$ mV m/N in a specific volume-fraction range have been discussed. The effect of electrical conductivity of the matrix on the pyroelectric activity of the composite has been studied, and examples of the volume-fraction behaviour of the pyroelectric coefficient $p_3^{*\sigma}$ have been analysed. The important argument that undoubtedly supports DEP as a high-effective method to manufacture the piezo-active composite consists in a good reproduction and stability of the properties of these materials.

A multiphysics finite element model has been put forward and applied to describe composite structure formation during DEP, where particles suspended in a dielectric Newtonian liquid between parallel-plate electrodes play an important role in forming long piezoelectric chains (0–3 and quasi 1–3 connectivity patterns). The performance of the model has been illustrated for a number of idealised starting configurations of the particles. The influence of the system of parameters concerned with the size, shape and spatial distribution of the particles has been studied to consider conditions for the formation of a desired composite structure. It becomes evident that a further optimisation of the effective electromechanical properties and related parameters of the composite would be impossible without knowledge of these parameters and ways for their prudent variation.

The important step at the study of the piezo-particulate composites is concerned with the prediction of their effective electromechanical properties. Hereby a number of models have been put forward at modelling and interpretation of the effective properties. Numerous examples of dependences of these properties on the volume fraction of FC, microgeometric characteristics of the composite and porosity in the polymer matrix have been analysed. A few micromechanical methods developed for piezo-active composites [5] have been used at the prediction of their effective properties. This prediction and careful analysis of the obtained results enable us to study the structure – property relations for a range of connectivity patterns, and this is effective for both the two- and three-component composites. We also show ways to improve the piezoelectric activity and sensitivity, figures of merit, piezoelectric anisotropy and other parameters [6, 7] that are important in various applications of piezoelectric materials.

In general, directions of the complex study on the piezo-particulate composites described in the present monograph are schematically pictured within the diagram in Fig. 6.1. We deal with the fundamental 'composition – structure – property' triangle that has been reinforced with the DEP method to manufacture novel composites and improve their effective physical properties. The main emphasis has been placed on the effect of DEP structuring on the composite microgeometry and properties, and ways to improve some parameters of the piezo-particulate composites have been also analysed. The effective properties and related parameters play a key role

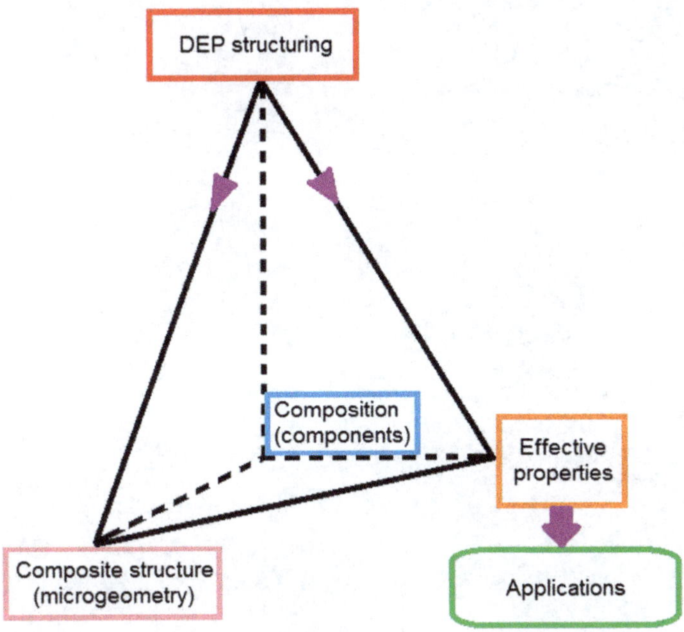

Fig. 6.1 Schematic of directions of study in the present monograph

in many piezo-technical, hydroacoustic, pyroelectric, energy-harvesting and other applications, and improving these properties promotes new effective applications of the composites.

References

1. S.A. Wilson, G.M. Maistros, R.W. Whatmore, Structure modification of 0–3 piezoelectric ceramic/polymer composites through dielectrophoresis. J. Phys. D Appl. Phys. **38**, 175–182 (2005)
2. H. Khanbareh, S. van der Zwaag, W. Groen, Effect of dielectrophoretic structuring on piezoelectric and pyroelectric properties of PT-epoxy composites. Smart Mater. Struct. **23**, 105030 (2014)
3. D.A. van den Ende, H.J. van de Wiel, W.A. Groen, S. van der Zwaag, Direct strain energy harvesting in automobile tires using piezoelectric PZT-polymer composites. Smart Mater. Struct. **21**, 015011 (2012)
4. H. Khanbareh, K. de Boom, S. van der Zwaag, W.A. Groen, Highly sensitive piezo particulate-polymer foam composites for robotic skin application. Ferroelectrics **515**, 25–33 (2017)
5. V.Yu. Topolov, C.R. Bowen, *Electromechanical Properties in Composites Based on Ferroelectrics* (Springer, London, 2009)
6. E.K. Akdogan, M. Allahverdi, A. Safari, Piezoelectric composites for sensor and actuator applications. IEEE Trans. Ultrason. Ferroelectr. Freq. Control **52**, 746–775 (2005)
7. C.R. Bowen, V.Yu. Topolov, H.A. Kim, *Modern Piezoelectric Energy-Harvesting Materials* (Springer International Publishing Switzerland, 2016)

Appendix A
List of Abbreviations

CAD	Computer-aided design
DEP	Dielectrophoresis
D-water	Demineralised water
ECF	Electromechanical coupling factor
FC	Ferroelectric ceramic
FEM	Finite element method
MCM	Mixed connectivity model
PCR	Piezoelectric ceramic from Rostov-on-Don (Russia)
PEO	Polyethylene oxide
PT	Ferroelectric $PbTiO_3$
PVDF	Polyvinylidene fluoride
PZT	Piezoelectric ceramic of the $Pb(Zr, Ti)O_3$ type
RPM	Revolution per minute
SEM	Scanning electronic microscope
ZTS	Piezoelectric ceramic of the $Pb(Zr, Ti)O_3$ type (Russia)

H. Khanbareh et al., *Piezo-Particulate Composites*, Springer Series in Materials Science 283, https://doi.org/10.1007/978-3-030-19204-4

Appendix B
Formulae of Polymer Components

In this Appendix, we give chemical formulae of some polymer components mentioned in the monograph.

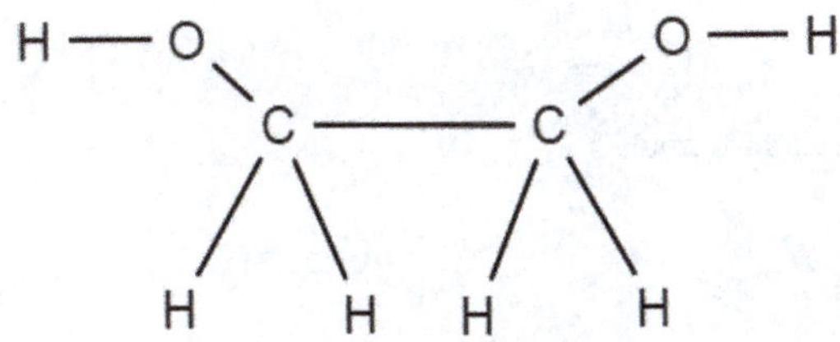

PEO, structural formula

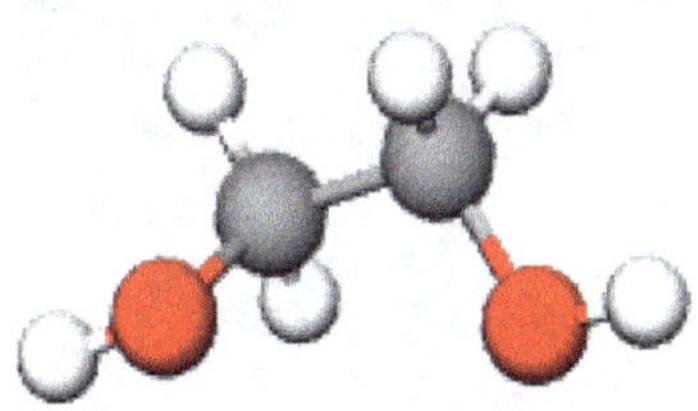

PEO, model [1]

H. Khanbareh et al., *Piezo-Particulate Composites*, Springer Series in Materials Science 283, https://doi.org/10.1007/978-3-030-19204-4

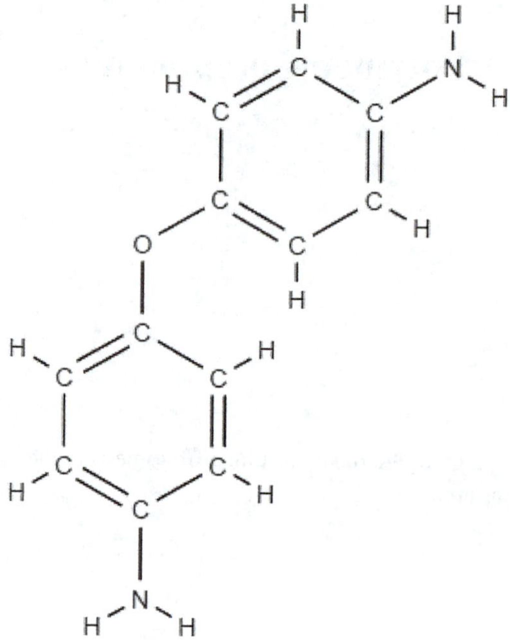

4,4'-oxydianiline, structural formula

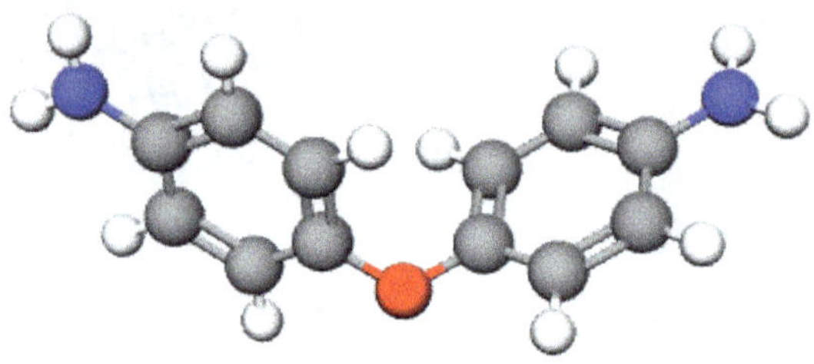

4,4'-oxydianiline, model

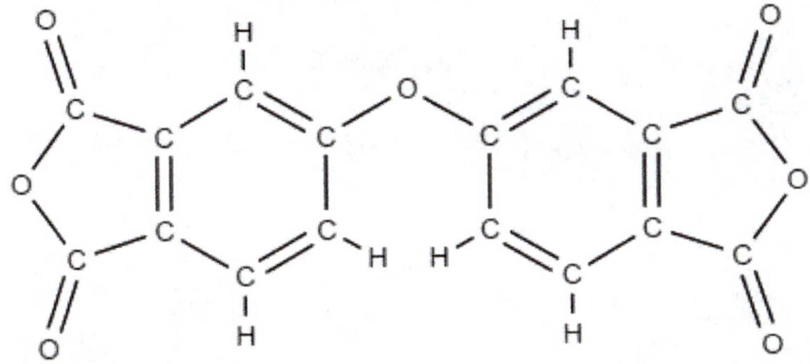

4,4'-oxydiphthalic anhydride, structural formula [2]

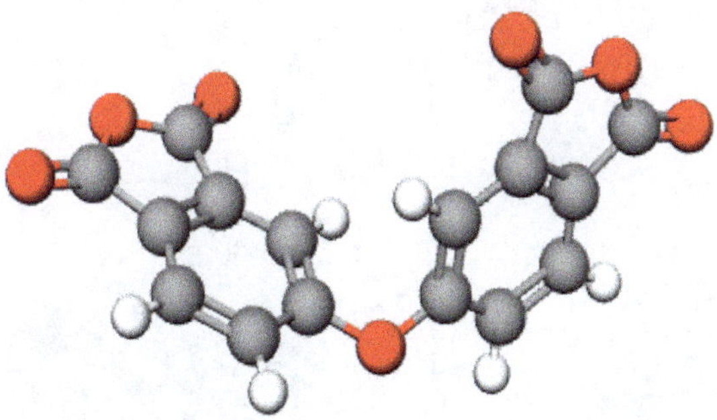

4,4'-oxydiphthalic anhydride, model

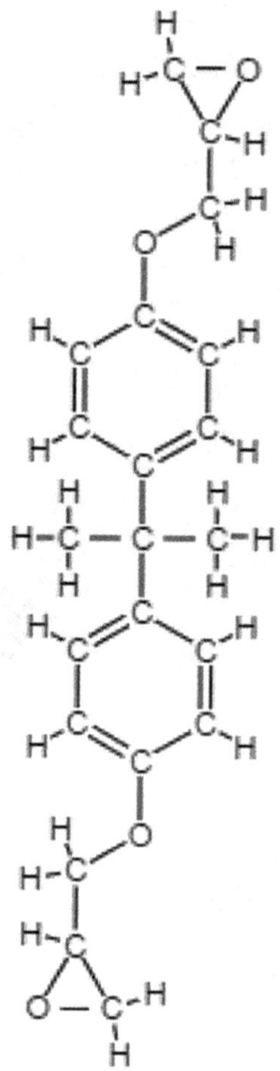

Diglycidyl ether of bisphenol-A, structural formula [2]

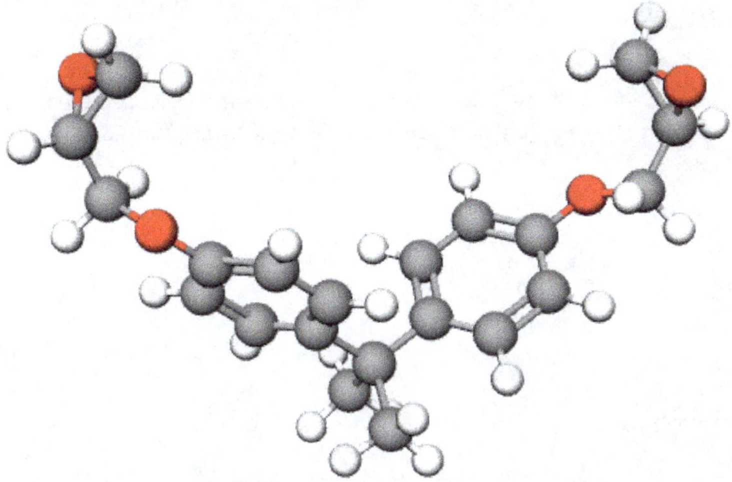

Diglycidyl ether of bisphenol A, model

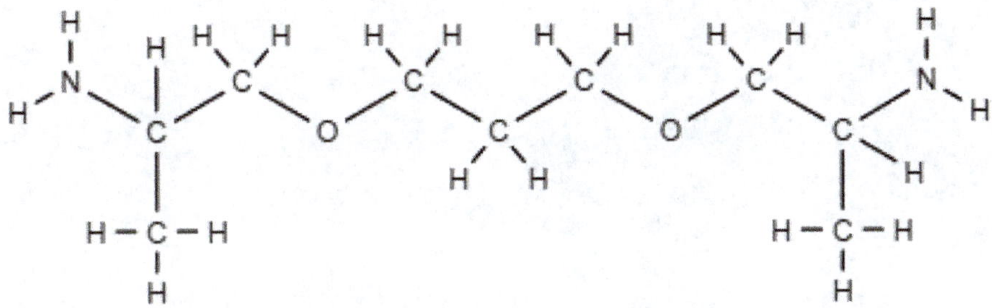

Polyoxypropyleneamine, structural formula

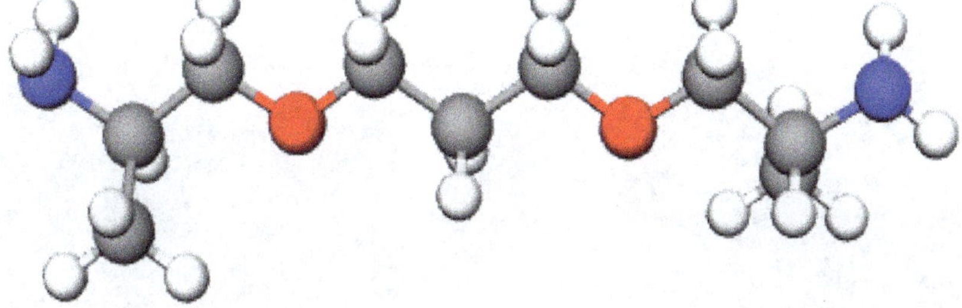

Polyoxypropyleneamine, model [2]

References

1. F.E. Bailey, J.V. Koleske, *Alkylene Oxides and Their Polymers* (Dekker, New York, 1990)
2. Physical properties of polymers, in *Handbook*, 2nd edn. ed. by J.E. Mark (Springer, New York, 2007)

Appendix C
Electromechanical Constants of Poled Ferroelectric Ceramics

FCs poled in the electric field $\mathbf{E}\|OX_3$ are characterised by ∞mm symmetry [1, 2]. In the rectangular co-ordinate system $(X_1X_2X_3)$, electromechanical properties of the FC are characterised by constants that are represented in the matrix form [1] as follows.

Elastic constants (elastic moduli c_{ab} or elastic compliances s_{ab}) are given by

$$\begin{pmatrix} w_{11} & w_{12} & w_{13} & 0 & 0 & 0 \\ w_{12} & w_{11} & w_{13} & 0 & 0 & 0 \\ w_{13} & w_{13} & w_{33} & 0 & 0 & 0 \\ 0 & 0 & 0 & w_{44} & 0 & 0 \\ 0 & 0 & 0 & 0 & w_{44} & 0 \\ 0 & 0 & 0 & 0 & 0 & w_{66} \end{pmatrix}$$

where $w = c^E$, c^D, s^E, or s^D.

Piezoelectric constants (piezoelectric coefficients d_{ij}, e_{ij}, g_{ij}, or h_{ij}) are given by

$$\begin{pmatrix} 0 & 0 & 0 & 0 & p_{15} & 0 \\ 0 & 0 & 0 & p_{15} & 0 & 0 \\ p_{31} & p_{31} & p_{33} & 0 & 0 & 0 \end{pmatrix}$$

where $p = d$, e, g, or h.

Dielectric constants (dielectric permittivities $\varepsilon^{\sigma}_{pp}$ or ε^{ξ}_{pp}) are given by

$$\begin{pmatrix} q_{11} & 0 & 0 \\ 0 & q_{11} & 0 \\ 0 & 0 & q_{33} \end{pmatrix}$$

where $q = \varepsilon^{\sigma}$ or ε^{ξ}.

H. Khanbareh et al., *Piezo-Particulate Composites*, Springer Series in Materials Science 283, https://doi.org/10.1007/978-3-030-19204-4

The elastic modulus c_{66} and elastic compliance s_{66} of every poled FC obey the relations $c_{66} = (c_{11} - c_{12})/2$ and $s_{66} = 2(s_{11} - s_{12})$, respectively [1]. Thus, the poled FC is characterised by five independent elastic constants, three independent piezoelectric constants, and two independent dielectric constants, so that the total number of the electromechanical constants in the full set equals 10.

Electromechanical constants of the poled FC are involved in (1.4)–(1.11). Among them, (1.4) and (1.5) or (1.6) and (1.7) are often used to describe the piezoelectric medium and its electromechanical properties [3–10]. The corresponding sets of electromechanical constants are c_{ab}^E, e_{ij} and ε_{pp}^ξ [in accordance with (1.4) and (1.5)], or s_{ab}^E, d_{ij} and ε_{pp}^σ [in accordance with (1.6) and (1.7)].

The full sets of electromechanical constants of the poled perovskite-type FCs are determined in experimental studies since the 1950s, and the number of such full sets known from numerous experimental data approaches 100 and remains the largest among the sets related to various structure types of widespread FCs. In Table 1.2, we listed the full sets of s_{ab}^E, d_{ij} and ε_{pp}^σ of the poled perovskite-type FCs. It should be added that the full sets of s_{ab}^E, d_{ij} and ε_{pp}^σ have been published in [3, 7–11]. The full sets of c_{ab}^E, e_{ij} and ε_{pp}^ξ can be found, for instance, in [4, 7, 11]. The aforementioned sets of electromechanical constants have been measured at room temperature [3, 4, 7–11].

References

1. I.S. Zheludev, *Physics of Crystalline Dielectrics*, vol. 2: Electrical Properties (Plenum, New York, 1971)
2. T. Ikeda, *Fundamentals of Piezoelectricity* (Oxford University Press, Oxford New York Toronto, 1990)
3. D.A. Berlincourt, D.R. Cerran, H. Jaffe, Piezoelectric and piezomagnetic materials and their function in transducers, in *Physical Acoustics. Principles and Methods*, Vol. 1: Methods and Devices. Pt A ed. by W. Mason. (Academic Press, New York London, 1964), pp. 169–270
4. L.P. Khoroshun, B.P. Maslov, P.V. Leshchenko, *Prediction of Effective Properties of Piezo-Active Composite Materials* (Naukova Dumka, Kiev, 1989) (in Russian)
5. K. Uchino, *Piezoelectric Actuators and Ultrasonic Motors* (Kluwer, Boston Dordrecht London, 1989)
6. A.V. Gorish, V.P. Dudkevich, M.F. Kupriyanov, A.E. Panich, A.V. Turik, *Piezoelectric Device-Making*, Vol. 1: Physics of Ferroelectric Ceramics (Radiotekhnika, Moscow, 1999) (in Russian)
7. V.Yu. Topolov, C.R. Bowen, *Electromechanical Properties in Composites Based on Ferroelectrics* (Springer, London, 2009)
8. V.Yu. Topolov, P. Bisegna, C.R. Bowen, *Piezo-Active Composites. Orientation Effects and Anisotropy Factors* (Springer, Berlin Heidelberg, 2014)

9. C.R. Bowen, V.Yu. Topolov, H.A. Kim, *Modern Piezoelectric Energy-Harvesting Materials* (Springer International Publishing Switzerland, 2016)
10. V.Yu. Topolov, C.R. Bowen, P. Bisegna, *Piezo-Active Composites. Microgeometry—Sensitivity Relations* (Springer International Publishing Switzerland, 2018)
11. A.Y. Dantsiger, O.N. Razumovskaya, L.A. Reznitchenko, L.D. Grineva, R.U. Devlikanova, S. I. Dudkina, S.V. Gavrilyatchenko, N.V. Dergunova, A.N. Klevtsov, *Highly Effective Piezoceramic Materials* (Handbook) (Kniga, Rostov-on-Don, 1994) (in Russian)

Index

H. Khanbareh et al., *Piezo-Particulate Composites*, Springer Series in Materials Science 283, https://doi.org/10.1007/978-3-030-19204-4

CPSIA information can be obtained
at www.ICGtesting.com
Printed in the USA
BVHW010217310519
549781BV00002B/3/P

9 783030 192037